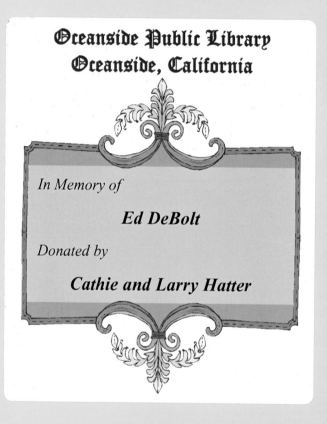

Oceanside Public Library
Oceanside, California

In Memory of

Ed DeBolt

Donated by

Cathie and Larry Hatter

GEDES
Nature's Treasures

Brad L. Cross and June Culp Zeitner

Gem Guides Book Co.
BALDWIN PARK, CALIFORNIA

Library of Congress Control No.: 2005936190

Softcover:
ISBN 10: 1-889786-32-2
ISBN 13: 978-1-889786-32-2

Hardcover
ISBN 10: 1-889786-34-9
ISBN 13: 978-1-889786-34-6

Book Layout and Design: Dianne Nelson, Shadow Canyon Graphics
Cover Design: Scott Roberts
Map Art: A-Frame Maps and Illustration

Photographs by Donnette Wagner and David Phelps or the authors, except as otherwise credited.

Published in the United States of America
Printed in China by TSE Worldwide Press, Inc.

Published by:
Gem Guides Book Co.
315 Cloverleaf Drive, Suite F
Baldwin Park, CA 91706
gembooks@aol.com
www.gemguidesbooks.com

Dedication

This book is dedicated to our very dear friend,
the late John Sinkankas, one of the finest men
and the most knowledgeable mineralogist and gemologist
we have known. Our times with him were
some of the best of our lives. His love, knowledge,
and enthusiasm for gems and minerals
will continue to influence all of us.

CONTENTS

FOREWORD

G o back in time if you will, millions of years ago, to a period when volcanoes and lava flows dominated the western United States, portions of Mexico, and vast regions of southern Brazil and Uruguay. It was a time when great oceans covered much of the central United States.

The conditions at that time favored rapid erosion of some areas and accelerated deposition of the remnants in others. The weathering away of mountains supplied large quantities of silt, clay, and sand, which were carried by streams to adjoining plains, basins, lakes, and oceans. These sediments would serve as the home for some of the beautiful geodes we collect today.

Life has inhabited the earth for over three billion years. However, some 360 to 325 million years ago, the lifeforms in the seas that covered what is now the Midwestern United States were vastly different from those of today. The first reptiles appeared during this time period and dinosaurs weren't even born. Sediment and calcium carbonate seashells accumulated at the bottom of the sea and over time were compressed to form mudstones, shales, and limestones that would serve as host to the beautiful sedimentary geodes we find today. Some of the simple (and often complex and bizarre) invertebrates would even later be preserved as sparkling geodes.

Millions of years passed to a time when lavas and vast ash flow tuffs poured out onto country of considerable topographic relief. They filled in irregularities of an old land surface and accumulated to thicknesses of 5,000 feet in places. This time period was when volcanic mountains such as Mt. Rainier, Mt. Hood, and other great giants formed. This explosive period in time provided the silica sources needed to form both the sedimentary and volcanic geodes discussed in this book.

Both on land and in the sea, the sight must have been truly spectacular. It was a time when the ground rumbled and the skies darkened with ash. The dinosaurs, flying reptiles, and toothed birds were gone. The specialized aquatic reptiles, such as plesiosaurs, ichthyosaurs, and mosasaurs were gone, and there were no longer any ammonites among marine invertebrates. On the whole, the invertebrates of this time period show great similarity to faunas living in the shallow sea bottom today.

The recorded history in the geodes and the rocks in which they are found provides us with a magnificent snap-shot in time of the rich history of the earth. Opening a geode and being the first to ever see the captivating hidden treasures inside is a thrill to anyone of any age. So come along with us as we explore the wonderful world of geodes from North and South America. We think you will find them as wonderful as we do.

ACKNOWLEDGEMENTS

*T*here are many that, from behind the scenes, have encouraged and supported this work. I am exceptionally fortunate to have *Donnette Wagner* as a professional photographer and friend. She very generously shared her time to assure the highest quality of photography.

Thank you *Hector and Jeannette Carrillo* (Gem Center USA) for your friendship, your generosity in providing specimens for photography, and allowing access to the various Mexican geode deposits. *Eugene Mueller* (The Gem Shop, Inc.), who took the time to carefully read what I wrote and made corrections, additions, suggestions, and improvements to my original words. *Tony Worth*, you provided seemingly unlimited belief in me and provided many hours of conversation on the various agate and geode deposits of Mexico. Thank you *Jonathan Parentice* for introducing me to Brazil. I have yet to find a trip that was as much fun. *Jeffrey Smith* provided a wealth of information and a number of photographs on the Trancas and San Marcos geodes as well as mid-continent geodes. Jeff, along with *Everett Harrington*, also provided a number of mid-continent geodes for my enjoyment. Without *Steve Wheeler*, I would have never been able to provide an adequate visualization of the Las Choyas geode mining activity. Thanks to *Cindy Brunell* (The Gem Shop, Inc.) for supplying photos of the wonderful polyhedroid specimens. *Vanderlei Dos Santos* and *Mano Czarnobay* of Aurora West were able to provide photos and a superb description of the Oco geode deposit in Rio Grande do Sul. *John Stockwell*, the leading authority on thundereggs, graciously reviewed our thundereggs chapter and provided a wealth of excellent suggestions. To *Karen, Benjamin,* and *Matthew* who tolerates me writing away when they really want me to put the pen down and pay attention to them. I cannot give back the lost evenings and weekends, but I can gratefully acknowledge your tolerance.

Finally, I'd like to thank *Gem Guides Book Company*. Although their official title is "publisher and editor," they are really so much more than that. For years they have graciously published many field guides which all of us greatly enjoy. They also combined the words of two authors and made them speak as one, caught errors, oversaw editing, production and art, and all aspects of this project.

— Brad L. Cross

ACKNOWLEDGEMENTS

Many thanks to *David Phelps* for the excellent pictures and for helping me examine geodes of all kinds. Thanks too to *Cecilia Gaston* for proofreading and retyping. Without the help of *Joe Nonast* I wouldn't have known about all the beauty of Englewood amethyst geodes. *Neal Larson* has been a great help with fine museum photos. *Edward Smith's* expertise on the Keokuks and Geode State Park has been notable. I'm so glad I found the perfect contact about the geodized fossils, *Margaret Kahr's* expertise is really appreciated. Was I surprised when mineralogist *Tom Loomis* found an important geode just minutes away from home. Thanks. *Vince Henderson* gave me the missing link geode from the Englewood. Much help! *Tim Bachand* was so kind to share his puzzling Colorado geodes with me. I am grateful to *Alta Vasey Morgan* and *Arlene Sohrweide* for those unforgettable geode hunts at Ballast Point. Special thanks to *Millie and Doug Heym* and *Layne Kennedy*. The Geological Surveys of Kansas, Indiana, Missouri, Kentucky, Tennessee, and Florida are exceptional resources. *May Hubbell* was very patient listening to all my frustrations, delights, and rewrites. I can't express my gratitude to *Dr. David F. Hess* enough. Thanks to him for the Geode Symposium book, the unpublished paper, and the fine pictures. Wasn't I lucky in knowing *Brad Cross*, fellow geode enthusiast and author so we could explore the geode world of the Americas together! What would the hobby do without *Gem Guides Books*?

— June Culp Zeitner

INTRODUCTION

Geodes are among the most fascinating and popular gemstones of the world. Hundreds of millions of these oddities of natural silica occur in both volcanic and sedimentary rocks many millions of years old in various parts of the world. When cut open, geodes can display an amazing wealth and variety of minerals and crystal arrangements. No two geodes are exactly alike; yet, all are thought to have been formed by similar processes, albeit processes which remain, in part, a tantalizing mystery.

Nothing can be more exciting than to find or purchase a geode, open it, and be the very first to view and discover the treasure found inside. In spite of their popularity, no book has previously been produced which is solely dedicated to geodes, although short articles in magazines have been published. The purpose of this book, therefore, is to introduce these beautiful works of nature to those people unaware of them and to present the latest information on the occurrence and formation of geodes.

The first section of the book defines and describes geodes that are found in igneous rocks and outlines their formation, distribution, and occurrence. We are taken to locations in Brazil and Mexico to visit the sites where the world's most popular geodes are found. The second section addresses a totally different type of geode – those provided to us from ancient seas. This section shares information on sedimentary geode formation and visits the most popular localities for these mysterious wonders.

Thundereggs and concretions, close cousins to geodes, are discussed in the third section. The final part of the book includes a glossary of terms as well as a list of frequently asked questions about geodes. A short bibliography is also included.

The authors were fortunate to have visited most of the geode localities in both North and South America where the study of agates and geodes have been one of their main interests. This book combines the vast knowledge of the authors in their respective areas of expertise. June Culp Zeitner shares over 60 years of study and field collecting from the geode localities in the United States, while Brad Cross brings extensive field mapping and collecting from the Mexican and South American geode localities.

ABOUT THE AUTHORS

*B*orn and raised in Corpus Christi, Texas, **BRAD CROSS** began serious field collecting at the age of twelve. At eighteen he won a national trophy for his lapidary work at the American Federation of Mineralogical Societies Show and, later received his degree in geology from the University of Texas at El Paso. He spent a number of years in Chihuahua, Mexico, mapping and mining Mexican agates, in the process becoming one of the foremost experts on agates. He is currently employed as a Senior Hydrogeologist with a private consulting firm in Austin, Texas.

His other published books include *Gem Trails of Texas* and *The Agates of Northern Mexico*. He is currently working on *Agates of the Americas*. He has also published numerous articles in *Rock and Gem*, one of the nation's leading magazines on the gem and mineral hobby.

*J*UNE CULP ZEITNER is considered one of the icons of the gem and mineral world. A professional writer since 1952, she has produced ten books plus thousands of magazine articles, columns, and book reviews. She has been on the staff of *Lapidary Journal*, *Earth Science*, and *Metal, Stone and Glass* as well as a contributor of many articles and illustrations to *Rock & Gem* and several other magazines and newspapers.

A featured lecturer and presenter at regional and national gem shows, she and her husband Albert traveled throughout the U.S. and to all the states of Mexico and the provinces of Lower Canada amassing a major display of self-collected minerals.

June started the national "State Stone" program and the Rockhound/Lapidary Hall of Fame. She was president of the Midwest Federation and chair of several AFMS committees, as well as Press Woman's Woman of Distinction and is on Northern University's Distinguished Alumni list. She is the 2005 recipient of the Carnegie Mineralogical Award for outstanding contributions in mineralogical preservation, conservation and education.

Part One

Gifts from Ancient Volcanoes

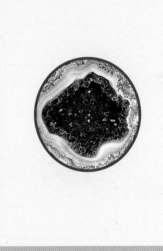

CHAPTER ONE

What Are Geodes?

— BRAD L. CROSS —

What are Geodes? Geodes are natural inorganic objects, most often chalcedony (pronounced kal-sid-knee) which are or have been hollow. Geodes are usually roughly spherical in shape and can occur in igneous or sedimentary rock. The interior may be lined with crystals, usually quartz, pointing toward the center. Quartz consists essentially of the chemical elements silicon and oxygen, but exhibits many varieties of color and form. Geodes are sometimes lined with chalcedony, a variety of quartz and, on occasion, closely related opal.

The term geode is derived from the Greek word *geoides*, which means "earth-like." The mysterious earth-shaped geodes have long challenged geologists to explain how they are formed. Geodes are a variable phenomenon and, therefore, many theories exist to explain how they are formed.

As the hollow interior of the geode is the most characteristic feature, it has become the custom to name any nodule that has a crystal-lined interior, a geode. If it shows concentric inward growth it is called a geode even though the crystals have filled the cavity. When the cavity becomes completely filled in, the formation is then properly called a *nodule*. The word "nodule" refers to the roundish formation of the rock and can be composed of any number of minerals – agate, quartz, or calcite. Sometimes a double name is used to give credit to both formational features as in agate-geode or septarian-geode. Most nodular structures may, on occasion, have internal openings, but are not designated as geodes because it is not their distinctive characteristic. Vein and other non-nodular formations may also contain hollow areas lined with crystals. These are properly referred to as vugs or crystal-lined vugs and not geodes. All geodes are nodules but not all nodules are geodes.

Geodes usually occupy the sites of former gas cavities (vesicles) in the volcanic rocks, basalt, rhyolite, and tuff. However, geodes also occur in various other environments, such as sedimentary rock, which occurs as shales formed by ancient oceans. The interiors of geodes are strikingly different from their exteriors and until the geode is opened, there is simply no way to predict what treasure will be found inside.

Geodes can also be found as loose stones, set free from volcanic lavas or massive limestones when the host rock is softened and crumbles through natural weathering. They are virtually unaffected by weathering and remain hard and resistant although their outer skins may be abraded by being tumbled about by the action of rain or a nearby stream.

The beautiful internal features can only be seen when the geode is cut or cracked open as discussed in later sections of this book. Geodes are commonly cut or broken into halves, revealing that they vary not only in the kind but also in the amount of quartz present. They may consist entirely of chalcedony or of chalcedony in association with one or more varieties of quartz such as rock crystal, amethyst, smoky quartz, agate, and jasper. Other minerals such as calcite (calcium carbonate), iron oxides, and manganese oxides are common associates of the aforementioned varieties of quartz. A minority of geodes does not have chalcedony or quartz, but can be barite, celestite, selenite, hematite, or goethite. Complex mineral associations and assemblages are not uncommon. It is the never-ending variety of mineral possibilities that make collecting geodes so appealing.

Some geodes are artificially dyed. Such examples do not find favor with the serious collector, who is usually able to discern natural material from dyed. Most electric blue, vibrant purple, bright green and flashy red specimens are artificially dyed.

The concentrated study of geodes is quite new. Not long ago there was no agreement on definition or origins. Even vugs were sometimes called geodes. As we learn more about fire born or water born geodes, more scientists and amateurs are becoming interested in the new puzzles and are giving them serious attention.

CHAPTER TWO

The Mysterious Formation of Geodes in Igneous Rocks

— BRAD L. CROSS —

Geodes have been collected and prized by man for thousands of years, yet until recently little was known about their origin and formation. During the last several hundred years many investigators both professional and amateur have sought the secret of geode formation; however, complete success has not been achieved. We also have no idea how long it actually takes for a geode to form.

The characteristic features of a geode found within a volcanic host rock are: an outer skin; a chalcedony, banded agate, or carbonate wall which grades inward into a well-defined euhedral quartz; and finally a complex late stage sequence of secondary minerals. This chapter will focus on geode formation in igneous rocks. The birth of mysterious sedimentary geodes will be discussed in Part II.

The formation of lava flows and the formation of geodes are not contemporaneous or even connected events. Most research has shown that geodes do not form in the final cooling stages of volcanic rocks as has often been assumed. In general, it appears that it is only after complete cooling of the volcanic flows that silica-bearing solutions penetrate the lavas and fill some of the vesicles with geode-forming material. The formation of geodes occurs over a range of low temperatures averaging 200° F or less.

Most theories accounting for the formation of geodes begin by explaining the formation of gas cavities in lavas. As molten lava is released onto the earth's surface, there is a drastic reduction in pressure. The falling pressure results in the formation of gas bubbles containing volatiles of various compositions, water vapor frequently accounting for 90 percent of the total gaseous content. The gas

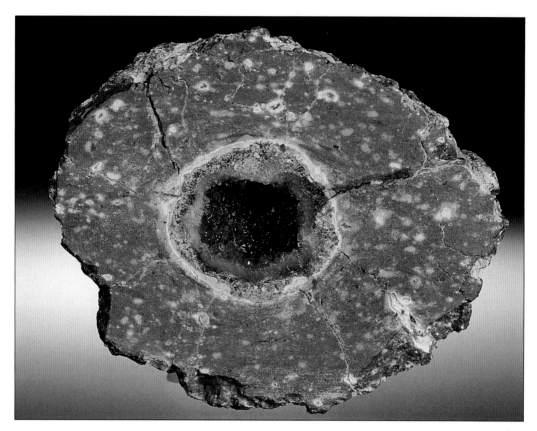

Geode in igneous host rock.
(PHOTO BY DONNETTE WAGNER; BRAD L. CROSS COLLECTION)

escapes quickly into the atmosphere, sometimes becoming trapped within the flows and held as bubbles within the lava as it continues to cool down, ultimately producing a cold lava that is full of holes (vesicles). The shape and orientation of the gas bubbles (and ultimately the geode) is reflective of the final flow direction.

The source of silica to form a geode is still one of the unsolved mysteries of geode formation. In many geode deposits, an overlying or underlying silica-rich geologic formation exists, possibly providing a source of leached silica solutions. The leached silica from the overlying or underlying formation is then re-deposited in the host vesicles. Many times the host rock itself contains silica minerals, as is the case in the Brazilian Amethyst geodes. There are several different views as to how the silica material gets into the empty gas cavities.

One theory holds that dilute solutions of silica enter the cavities through all points in the cavity wall and begin precipitating thin deposits of silica against the wall. This silica precipitate later crystallizes as chalcedony. As solutions become depleted of silica over time, quartz crystals begin lining the interior of the geode. Variations in the chemistry of the solution determine what variety of quartz will form. Trace amounts of iron within the silica solution may produce amethyst while a solution containing aluminum could provide smoky quartz.

A second theory, however, cannot accept that solutions can enter or escape from cavities once the early-formed layers have coated the inner walls and crystallized solidly against them. Supporters of this theory suggest instead that dense silica gels are formed and remain inside the cavities. Within the containment of each cavity, various chemical phases are segregated by diffusion and eventually crystallize as chalcedony or quartz, or both.

A third theory begins with an empty gas cavity. The cavity is usually lined with a mineral skin consisting of one or more silicate minerals such as celadonite, chlorite, or saponite. The skin is formed by meteoric or ground waters that leach and decompose constituents from the host rock and eventually deposit them in the empty gas cavities where they crystallize as a mineral skin. This coating may appear as the dark green mineral celadonite or perhaps other chlorite-rich minerals.

The vesicle is next penetrated by silica-rich solutions that wholly or partially fill the void with a dense silica gel. The origin for these solutions is again meteoric, or possibly ground water, which has leached silica from an overlying or perhaps underlying geologic formation. It has also been suggested that the source for water could be small alkaline lakes located on the surface above the vesicles. The defined surface water bodies would certainly explain why some geode deposits are isolated to specific areas within a region. These waters leach the silica and other constituents from rock-forming minerals and deposit them in the empty vesicles as a dense plastic-like gel. In some geodes the skin may rupture and fragments of it may hang down into the gel. Features such as this are quite common in the large spectacular amethyst geodes from Brazil.

The gel is now contained in its own unique space and increases in density until it reaches a crystalline state. During this process, a separation or segregation of its constituents into layers through chemical differentiation and diffusion begins taking place. It is these different layers that will crystallize as layers of chalcedony, common quartz (also known as rock crystal), amethyst, or smoky

Simple Diagramatic Sketch of Geode Formation
(MODIFIED AFTER MACPHERSON, 1989)

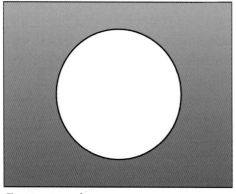

Empty gas cavity.

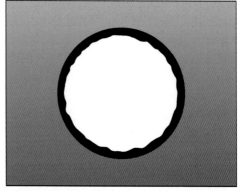

Skin coating wall of gas cavity.

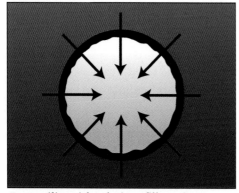

Dense silica-rich solutions fills cavity.

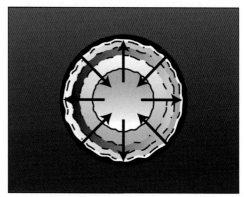

Separation of solution into hydrous and anhydrous layers begin.

Separation continues and layering becomes distinct. Variations in silica and water content as well as temperature will affect when layers of agate or chalcedony end and formation of quartz crystals begin. The size of the geodes internal cavity is proportional to both the amount of available silica and the amount of water in the silica.

quartz. Each individual layer contains its own composition of silica, water, and impurities to provide a geode unlike any other. Variations in silica and water content, as well as temperature, will affect when layers of agate or chalcedony end and formation of quartz crystals begin.

Whatever theory you subscribe to, the nature of all the forces remains unknown, yet we should not be deterred from recognizing that such forces exist. A problem that has been around for hundreds of years is not going to be solved overnight. Every geode is one of a kind even though the rock environment within that area is the same with respect to pressure, temperature, and general chemical composition. Geodes of totally different kinds with totally different features can grow side by side, each in its own unique abode. This is why enthusiastic collectors call them "nature's surprise packages."

Mexican "Coconut" Geode, Las Choyas, Chih., Mexico
(PHOTO BY DONNETTE WAGNER; BRAD L. CROSS COLLECTION)

CHAPTER THREE

Inedible Coconuts
Mexican "Coconut" Geodes

— BRAD L. CROSS —

I will never forget my first trip to the coconut geode deposit and the agony of that sunny El Paso day, now so long ago. I stood in the shade cast by a narrow lean-to roof. There I stood awaiting the end of a hot day, admiring a long row of large open boxes filled with recently mined rams horn selenite. Any moment my friend Gerald Berry and I would depart for geode and agate country, taking me on my first trip to a land that I dearly love.

It was the summer of 1977 and I was a sophomore at the University of Texas at El Paso. If freedom is, as Kris Kristofferson wrote and sang, just nothing left to lose, I was certainly free then. We were travelling in a gutted van, unlike the fancy luxury vehicles of today. Gerald closed up his mineral shop and, as the sun began to set, we departed for the Las Choyas coconut geode deposit.

A rockhunting trip to this remote region of Mexico had long been a dream of mine. The coconut geodes and the world's most colorful agates had always intrigued me. What was the area like? Were agates and geodes laying on the surface simply awaiting someone to pick them up? Was the rugged area as romantic as I had imagined in my dreams as a child? To me it was a land of mystery – in its history, people, geology, and geography.

I wasn't bothered by the lack of facilities in the region – no electricity, no water. In my mind, I was an adventurer, a pioneer. Of course we would camp out; no tourist reservations for me, even if there had been accommodations. I had my army surplus backpack and canteen.

Villa Ahumada was the jumping off point then, as it is today. We filled the van with gas and stopped at a corner meat market to buy some warm asadero

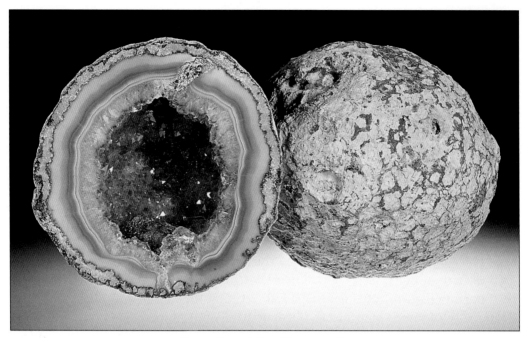

Cut and rough Las Choyas geode.
(Photo by Donnette Wagner; Brad L. Cross Collection)

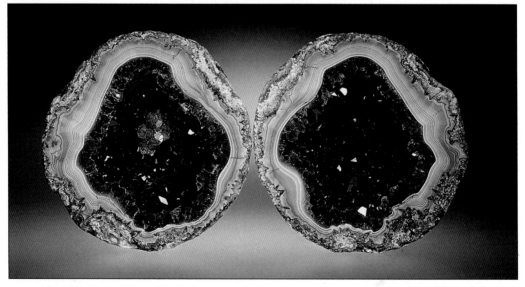

Magnificent deep purple amethyst crystals line this coconut geode. Amethyst this dark is extremely rare. (Photo by Donnette Wagner; Brad L. Cross Collection)

burritos. At the time, I was ignorant of much of the area's history and oblivious to the remains of Indian occupation that predated the first western settlement. Little did I realize that we were parked right in the middle of the *Camino Real* – in those days two ruts worn deep around dunes, through arroyos, and up mesas, connecting the sleeping civilization of New Mexico to the capital of Mexico. It was a trail of eighteen hundred miles, and as far removed in advancement. That evening was when I saw Mexico. She captured my soul and she still has it. I walk into the geode and agate country and I am home.

The "coconut" geodes from Las Choyas, Chihuahua, Mexico, (also known as El Mesteño, a name given to a nearby large rhyolitic intrusion) are perhaps the most popular of all geodes, rivaled only by the monstrous amethyst geodes of southern Brazil. They are sold at all rock shows, rock shops, and many gift shops throughout the world. The remote locality is about 22 miles east-northeast of Laguna Enciñillas, Chihuahua, or 143 miles south-southeast of El Paso, Texas. These quartz geodes are mined from a two square-mile area and have constituted a multi-million dollar business. No one knows exactly how many pounds have been produced because production records have never been kept. Geodes from this location are easily identifiable by their near-perfect spherical shape. They occur in a feldspar-rich ash flow tuff and the geodes when brought to the surface appear white from the clinging fragments of the volcanic ash in which they were embedded.

The geodes from this Mexican locality are many times termed "coconuts." This name was given to the geodes not because of their similar appearance to the fruit, but instead they were named after a Mexican woman who occasionally peddled the geodes. Reportedly, she was a bit crazy and was nicknamed "La Cocona," thus the locals termed her geodes "coconuts." Chances are that any Mexican geode or coconut that the reader has seen in rock shops or shows originated from *La Estrella, La Otra Estrella, La Animosa, El San Antonio, La Morenita, La Paty, Esperanza, Santa Rosa,* or *El Mesteño* claims at this location.

Many other varieties of geodes and thundereggs have come from northern Mexico and are described later within this book.

Coconuts are located within the Chihuahuan Desert – the largest and most remote desert region in North America. At 175,000 square miles, it is larger than the entire state of California. Most of it is located in the states of Chihuahua and Coahuila in Mexico, but fingers of the Chihuahuan reach up into eastern Arizona, southern New Mexico, through the Trans-Pecos of West Texas, and down into the Mexican states of Durango, Nuevo Leon, Zacatecas and San Luis Potosi.

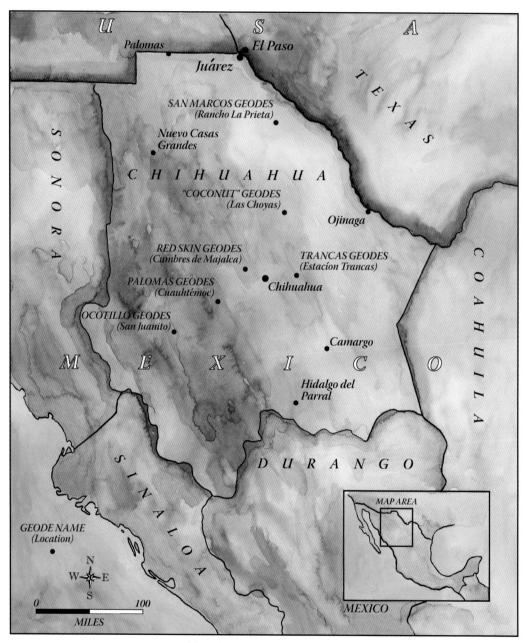

General location map of Chihuahua, Mexico
(MAP BY MARK LESH)

The area where Mexican coconuts are found is a remote region of the Chihuahuan desert. Rainfall is less than nine inches per year and the area is sparsely populated with creosote bush and other desert scrub.

The craggy and majestic mountains surrounding the coconut geode deposit are igneous rocks, mostly andesites, tuffs, and basalts. All are underlain by Cretaceous limestone. This area is marked by desert climate and vegetation and is situated in a scarce rainfall province that has an average precipitation of nine inches. Aridity imposes a major constraint on vegetation and soils in the area. Various scrub species and semi succulents such as Lechuguilla, Sotol, and Yucca dominate the coconut country. Creosotebush is a prominent element of the desert scrub, continuing to tinge the air with its peculiar scent following a rainstorm. Desert grasslands are scattered throughout the region.

The geode-producing region is an environment demanding the utmost in tenacity of spirit, courage, and strength of fiber of mortal man. It is a somber, silent realm of tawny plains with little water and shade or high rugged mountains cut by vast canyons. Sudden storms of sand or rain come violently then depart. The sun blisters and cracks, the wind sears, and in time the land subtly instills fear and disloyalty.

The essence of geode country is found in quiet haciendas spaced along the watercourses. Here are random groupings of dark-colored mud houses forming a backdrop for a simple, whitewashed chapel. Here are unmortared fences of piled boulders following up over ridges and down gravelly draws. Here are chickens, pigs, and burros grazing by the roadside. This is Chihuahua!

A 'brown rimmed' coconut geode.
(Photo by Donnette Wagner; Brad L. Cross Collection)

Mineralogy

"Coconut" geodes usually have an outer wall of variable thickness consisting of blue-gray chalcedony. However, the walls of some coconuts consist of a dark brown iron carbonate, a mineral known as siderite, and are termed "brown rimmed coconuts." The wall grades inward into well-defined crystalline quartz. Finally, there is a complex late-stage sequence of minerals, including, carbonates, manganese oxides, and iron oxides and hydroxides in the centers of many of the geodes.

The mineralogy of these geodes has been studied extensively and has revealed a total of at least eighteen different minerals. Seven of these are manganese oxides, some of which are considered quite rare. These secondary minerals occur in a great variety of colors and forms. Many are extremely small, usually less than one millimeter long, and to the unaided eye they often appear as small black specks on or in the quartz and calcite crystals that line the interior of the geodes. Even the solid specimens, sold as "solids," are completely filled with banded agate or common quartz, and contain trace amounts of manganese or iron oxides.

Pale amethyst coconut geode with secondary black todorokite crystals scattered throughout.
(PHOTO BY DONNETTE WAGNER; BRAD L. CROSS COLLECTION)

Some of the geodes contain large plates of white calcite that criss-cross the interior and are reminiscent of a modern day sculpture. Other secondary minerals may appear as "spots of dirt" on the sparkling crystals. But don't scrub that dirt out! Take time to inspect the geodes with a hand lens or microscope. There is a whole new world waiting to be discovered. Under the microscope these specks are actually whiskers and clusters of perfectly terminated iron and manganese oxides. They may be combined with fancy varieties of quartz, providing a complex and outstanding mineral assemblage, which rivals any known to man. Sometimes the freshly opened geode will provide a cavity filled with mordenite – an unusual white and spongy zeolite mineral that is reminiscent of gooey mold. The Mexican coconuts are indeed a haven for the mircromount mineral collector!

Minerals removed from the geodes have been identified by their optical properties, X-ray and electron diffraction, and by elemental analysis with an electron microprobe X-ray analyzer. The minerals identified to date are listed below in frequency of occurrence.

Some coconut geodes contain plates of randomly oriented calcite and are reminiscent of a modern day sculpture. (PHOTO BY DONNETTE WAGNER; BRAD L. CROSS COLLECTION)

Mordenite-filled Las Choyas geode.
(PHOTO BY KIM JONES; BRAD L. CROSS COLLECTION)

Quartz (SiO_2)

The primary mineral found in Las Choyas geodes is quartz, forming a zone on the chalcedony or siderite outer shell. The apex of the pyramid generally points to the center of the geode. Frequency of occurrence is approximately 90%. Colorless quartz often appears dark due to transmission of the colors of underlying minerals. Inclusions in the primary quartz are common. Included minerals are goethite, ramsdellite, and hematite. Smoky and amethystine varieties occur in approximately 20% of the geodes.

Quartz also occurs as a secondary mineral crystallizing over the primary quartz, is randomly oriented, and doubly terminated. Secondary quartz is also observed crystallized on other secondary minerals such as goethite and calcite.

Chalcedony (var. Agate) (SiO_2)

Found directly beneath the rough exterior shell, the milky white to pale blue-gray banded agate occurs in approximately 75% of the Las Choyas geodes. The thickness is variable. Approximately 25% fluoresce bright green in short wave UV light.

Siderite ($FeCO_3$)

Occurs in approximately 25% of the Las Choyas geodes as an outer wall. The thickness is variable.

Calcite ($CaCO_3$)

Calcite is the second most common mineral found in Las Choyas geodes next to quartz. Dogtooth (scalenohedral) and thick colorless to white plates are the most common forms. It is present in approximately 50% of the hollow geodes and can occupy up to 90% of the internal cavity. The dogtooth habit is frequently included with an unidentified brown mineral resulting in a brown to light tan apparent coloration. Two generations of calcite crystallization appear to be present in the Las Choyas geodes.

Hematite (Fe_2O_3)

Hematite occurs as black, red, orange, or brown hemispheres, spheres, and rosettes in 50 to 75% of Las Choyas geodes. The mineral is often observed as inclusions in primary quartz. Many of the spheres have a prominent basket-weave texture.

Goethite ($Fe_2O_3 \cdot H_2O$)

Goethite is the most common accessory mineral found, and occurs in approximately 50 to 75% of the Las Choyas geodes. It is found as black, red, orange, and brown bladed crystals. It often occurs in radiating or sheath-like groups with chisel-like terminations and as globular aggregates. The two morphologies are often found together in the same geode. In the sequence of mineral deposition, the goethite appears to have formed towards the end of the primary quartz crystallization and continued after the primary quartz crystallization had ceased.

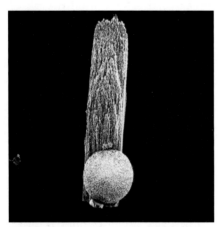

Hematite hemisphere growing at the base of a stack of goethite blades. Magnification x40. (PHOTO COURTESY OF ROBERT B. FINKELMAN)

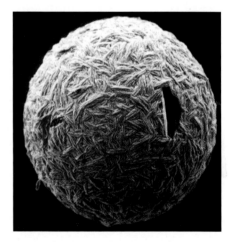

Hematite sphere with prominent basket-weave texture. The sphere grew around a blade of goethite. The hole through which the blade exited is seen on the right side of the sphere. Magnification x200. (PHOTO COURTESY OF ROBERT B. FINKELMAN)

Todorokite [(Mn,Ca,Mg) $Mn_3O_7 \cdot H_2O$)]

Todorokite is light brown to black fibrous overgrowth on other minerals. It is commonly found as dense mats composed of long fibers (up to 4 mm). Todorokite occurs in about 30% of the Chihuahua geodes, but volumetrically it is the most abundant manganese mineral.

Todorokite and late stage quartz precipitated simultaneously as the last crystals in some geodes, thus forming quartz crystals impregnated with dispersed strands of fine todorokite fibers.

Ramsdellite (MnO_2)

Ramsdellite occurs as small black crystals, many times on goethite blades. At least six different habits have been observed. The most common habit, generally consisting of radiating crystal aggregates to about 3 mm in length. The crystal terminations are generally more pointed than those of goethite. Ramsdellite occurs in approximately 25% of the geodes. The morphology and precession X-ray patterns suggest the ramsdellite in Las Choyas geodes may be pseudomorphous after groutite.

Acicular goethite. Magnification x150.
(PHOTO COURTESY OF ROBERT B. FINKELMAN)

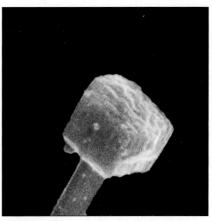

Platy goethite on the tip of acicular goethite. Magnification x15,000.
(PHOTO COURTESY OF ROBERT B. FINKELMAN)

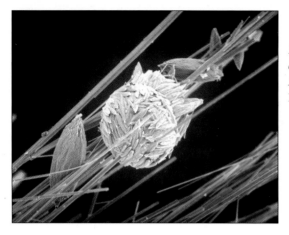

Ellipsoidal quartz crystals covering a chalcedony spherulite on mordenite fibers. Magnification x300. (PHOTO COURTESY OF ROBERT B. FINKELMAN)

Hematite rosette on a goethite blade. Magnification x300. (PHOTO COURTESY OF ROBERT B. FINKELMAN)

Acicular goethite with platy goethite crystals at the end of the needles. Magnification x150. (PHOTO COURTESY OF ROBERT B. FINKELMAN)

Pyrolusite (MnO_2)

Black to brown pyrolusite dendrites occurs between agate bands in the Las Choyas geodes. The dendrites are curved parallel to the circumference of the geode, inferring that the manganese formed on the surface of an agate band and was then covered by subsequent crystallization of chalcedony. The dendrites can be observed in less than 20% of the specimens.

Pyrolusite is also found in association with ramsdellite and is believed to be an alteration product of that mineral.

Mordenite [$(Ca, Na_2, K_2)\, Al_2Si_{10}O_{24} \cdot 7H_2O$]

Mordenite occurs as masses of fine white cottony material, many times completely filling the hollow interior. Habits include spongy fibrous mats and as clusters of long slender fibers. Mordenite occurs in less than 5% of Las Choyas geodes.

Gypsum ($CaSO_4 \cdot 2H_2O$)

Gypsum is a fairly rare mineral occurring in less than 1% of the geodes. It occurs as very small, colorless, tabular crystals.

Opal ($SiO_2 \cdot nH_2O$)

Occurring in a colorless to white botryoidal habit, opal is sometimes found on blades of goethite and is many times covered by a black film of manganese oxide. Opal is found in less than 1% of the geodes.

Kaolinite [$Al_2Si_2O_5\,(OH)_4$]

Kaolinite can only be observed in Las Choyas geodes using an electron microscope. Occurring as white hexagonal plates, the frequency of occurrence is less than 1%.

Apatite (var. Dahllite) [$Ca_{10}\,(PO_4)_6(CO_3) \cdot H_2O$]

Dahllite occurs as colorless to pale green hexagonal plates usually found on hemispheres of hematite. Frequency of occurrence is less than 1%.

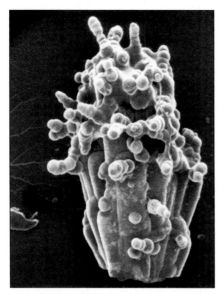

(Left): Goethite blades covered by a botryoidal amorphous silica. Magnification x50.
(PHOTO COURTESY OF ROBERT B. FINKELMAN)

(Right): A quartz crystal growing around a hematite covered goethite needle. Magnification x900.
(PHOTO COURTESY OF ROBERT B. FINKELMAN)

(Left): Randomly oriented quartz crystals. Magnification x80.
(PHOTO COURTESY OF ROBERT B. FINKELMAN)

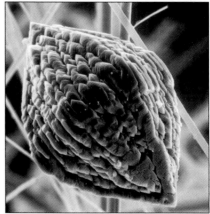

(Right): Quartz dipyramid on mordenite fibers. Magnification x1,500.
(PHOTO COURTESY OF ROBERT B. FINKELMAN)

24

Cluster of radiating quartz crystals on mordenite fibers. Magnification x260. (PHOTO COURTESY OF ROBERT B. FINKELMAN)

Hematite rosettes, goethite needles, and euhedral quartz crystals all growing on an earlier generation of goethite. Magnification x100. (PHOTO COURTESY OF ROBERT B. FINKELMAN)

Ellipsoidal quartz crystal on mordenite fibers. Magnification x300. (PHOTO COURTESY OF ROBERT B. FINKELMAN)

Beidellite $[(NaCaO_{.5})_{0.3}Al_2(Si,Al)_4O_{10}(OH)_2 \cdot nH_2O]$

Light gray in color, beidellite occurs as fine stubby fibers in less than 1% of the Las Choyas geodes.

Birnessite $[(NaCa)Mn_7O_{14} \cdot 3H_2O]$

Black birnessite occurs as a granular fracture filling in calcite. Frequency of occurrence is less than 1%.

Rancieite $[(Ca,Mn)Mn_4O_9 \cdot 3H_2O]$

Generally associated with calcite, rancieite occurs as black irregular masses, dendrites, and hexagonal plates in less than 1% of Las Choyas geodes. Rancieite is generally associated with secondary calcite.

Cryptomelane $[K(Mn^{2+},Mn^{4+})_8O_{16}]$

Black fibrous tufts of cryptomelane occur in less than 1% of Las Choyas geodes.

Hollandite $(BaMn_8O_{16})$

Hollandite occurs in less than 1% of Las Choyas geodes as black, slender needles.

Most hollow geodes contain at least one manganese oxide and many have more than four. This list of minerals is not considered to be complete. Undoubtedly, other minerals will be added to the list as investigations continue.

HOW ARE COCONUTS FORMED?

The coconuts, like many other geode deposits in Northern Mexico, form in what is known as an ash flow tuff. What is an ash flow tuff? It is a volcanic rock of andesitic (medium to dark in color, containing 54 to 62% silica, and moderate amounts of iron and magnesium) to rhyolitic (light in color, containing 69% or

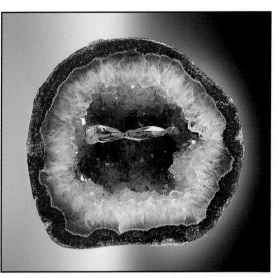

An amethyst coconut geode with a lustrous, water-clear sceptered quartz crystal growing up from the floor of the geode. (PHOTO BY DONNETTE WAGNER; BRAD L. CROSS COLLECTION)

An extremely rare phenomenon with two secondary calcite crystals, each of which has grown from the opposite side of the geode and along a single goethite blade. (PHOTO BY DONNETTE WAGNER; BRAD L. CROSS COLLECTION)

more silica, and rich in potassium and sodium) composition that has formed in an unusual way. It usually contains some large crystals, but the major component is particles of volcanic glass. It forms when gas dissolved in the lava suddenly boils as the roof of the magma chamber collapses. This causes the lava that has been slowly crystallizing to explode. The pressure blows some of the molten rock into tiny fragments that cool rapidly to fine dust or ash. This collapse of an eruption column produces a hot, dense, laminar flow of volcanic debris. Fluidization (an upward movement of gas or water causing the particles to behave as a fluid) is brought about by the expansions of gases exsolved from the magma and of air caught up in the advancing flow.

The flows can travel for several hundred miles and are commonly 50 feet thick or more. Pumice and glass fragments become flattened and stretched during flow and compaction of ash flow deposits. This compaction is termed *welding* and occurs after deposition as the hot plastic ash particles in the central part of the ash flow typically become welded together to form a dense rock. As a result of the more rapid heat loss, the lower and upper parts of an ash flow deposit are usually non-welded and thus have a higher porosity. It is usually in these non-welded "cooling units" that we find the geodes.

Formation of the Las Choyas geodes begins when the host ash flow tuff is very hot and plastic. The basal portion, which has come into contact with the colder ground, chills rapidly and forms a glass or basal vitrophyre. The unit is highly charged with gas. When it cools, the glassy texture converts to a crystalline texture (devitrification) with round aggregates of feldspar and kaolinite. These aggregates are known as spherulites. As the crystallization of the glass proceeds, the gas held in solution in the glass exsolves and collects around the centers of crystallization (the spherulites).

As gas is released and concentrates in the interior of the spherulite as a growing bubble, the spherulite expands. These expanding spherulites will serve as the void within which geodes will later form. The number of geodes within a zone is sometimes so great that the force exerted by adjacent geodes often produces irregularities in shape. Under these circumstances, it is not uncommon for two or more geodes to be cemented together in a single mass. When the perimeter of two spherulites touch each other, the adjoined and resulting geodes are referred to as a "double." Three adjoined geodes are referred to as a "triple."

We know that at Las Choyas there was a time lapse of perhaps several million years during which ground water circulated through the permeable tuff and deposited silt and clay in many of the cavities. This gave rise to what are known as "mud balls" which exhibit a wide range of fascinating sedimentary structures such as graded bedding, cross bedding, and slump features.

The sedimentation process may have lasted until nearby rhyolite domes intruded the area. The heat from these intrusions warmed the ground water and caused the start of a convective process. The heated water circulated through the tuff, breaking down the feldspar within the tuff and leaching it of its silica.

The highly saturated silica solution crystallized rapidly and the formation of chalcedony began taking place. This usually represents the outer rims of the geodes. By the time the groundwater was depleted of silica and the rate of crystallization slowed, the individual crystals of quartz were forming on the interior of the cavities.

Oxygen isotope studies conducted by Keller (1977) to determine the temperatures of silica deposition indicate that the chalcedony rinds of the geodes were deposited at temperatures ranging from about 118° to 138° F, and the quartz on the interiors of the geodes formed at significantly higher temperatures, between 156° and 174° F. These results suggest that the groundwaters were heating up during mineralization and depletion of the silica was occurring.

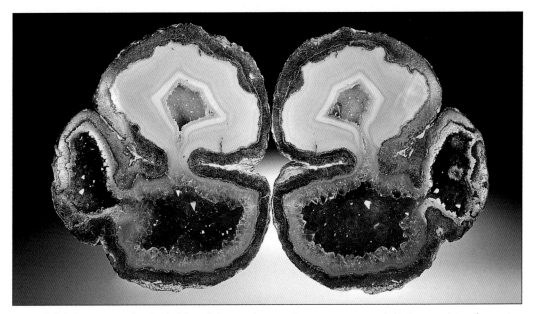

A 'triple' coconut geode. Each lobe of the geode contains a separate and distinct variety of quartz. The first lobe contains only amethyst, the second lobe contains smoky quartz, and the third lobe consists of white shadow agate. All three geodes formed side by side under identical environmental conditions. (PHOTO BY DONNETTE WAGNER; BRAD L. CROSS COLLECTION)

A 'mud ball' which suggests there was a time lapse of perhaps several million years between the time the cavity formed and mineralization took place. (PHOTO BY DONNETTE WAGNER; BRAD L. CROSS COLLECTION)

Some coconut geodes have been completely faulted, offset, and cemented. Features such as this are due to Basin and Range faulting some 10 to 15 million years ago. (Photo by Donnette Wagner; Brad L. Cross Collection)

MINING OPERATIONS

A mine-run or average geode from Las Choyas ranges from 2.0 to 4.5 inches in diameter and is 20% hollow. They are found within an ash flow tuff that has been altered to a grayish-white montmorillonite clay. The geode-producing unit varies from one foot to five feet in thickness. The unit, known locally as the Liebres Formation, is about 44 million years old. The producing unit dips west-ward at seven to ten degrees. Therefore, the geode-producing horizon is at greater and greater depths as one goes westward, away from its outcrop on the slope where the geodes were first discovered.

The coconut geode mining area. The large piles of white material is altered ash flow tuff from underground mining activities. (Brad L. Cross Collection)

Mine shafts are currently drilled using an old water well drilling rig.

A headframe is placed over the mining shaft – a three foot diameter hole extending downward to 100 feet.

The mines in the Las Choyas area average depths of 100 feet and in certain areas can reach 125 feet. This means that some 100 feet of solid volcanic overburden must be removed to reach the geodes. Roughly three-foot diameter shafts are dug through tenacious ash flow tuff. Once the geode-producing unit is reached, tunnels are constructed in the highly altered tuff, following the pay zone. New shafts are constructed every 50 to 100 feet depending upon what the miners feel is "safe." Many of the tunnels connect with each other while some are backfilled to help dispose of waste rock and muck as well as provide structural integrity.

Not many years ago, the vertical shafts had to be completely dug by hand. Imagine hand digging a three-foot diameter hole twenty to sixty feet deep in tenacious volcanic rock. Needless to say, this was not only time-consuming, but

Schematic View of Coconut Geode Mining

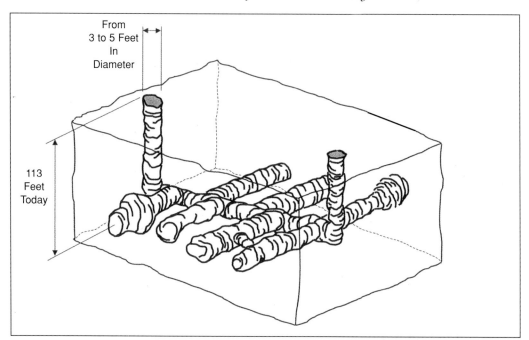

From
3 to 5 Feet
In
Diameter

113
Feet
Today

(ILLUSTRATION BY STEVE WHEELER)

A diesel-powered generator helps provide the power to remove barrels of muck as well as coconut geodes from deep below the surface.

In years past, small wood 'cigueñas' were utilized to provide access to the mine.

32

The author underground. The tunnels dip at about 10°, making the underground climb quite difficult, especially in an environment with little air. (PHOTO BY STEVE WHEELER)

One-hundred feet below the surface, a maze of tunnels can be found, extending in all directions, following the geode-producing horizon.

also resulted in a revenue loss to both the claim owner and the miner. (Miners are paid based upon pounds of geodes produced.) Rotary drilling equipment is now used to drill a vertical pilot hole that is later reamed with a larger diameter bit. Miners then widen the shaft by hand to almost three feet in diameter. What now takes no more than several weeks once took miners several months.

It is indeed quite an experience to slip down into a one-hundred-foot-deep shaft, hand over hand on a one-inch rope into a dark maze of tunnels. If you aren't claustrophobic before slipping down into this tight hole, you suddenly realize that anything is possible. Crawling through the low tunnels, it's not uncommon to hear nearby dynamite explosions that rattle the earth around you. A damp clay smell permeates the mine air and without light, you can't see your hand in front of your face. Before the advent of solar powered lights, smelly carbide lamps would tinge the thin, muggy air. Today low-watt bulbs are powered by a solar generator located on the surface near the mineshaft. Through the use of a pulley system, this same generator assists the miners in lifting both the waste rock and geodes to the surface. Tunnel heights range from several feet up to six feet.

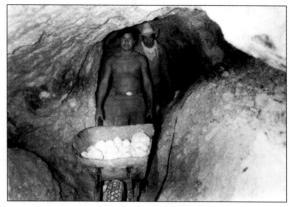

Tarahumara miners transport geodes underground where the nodules will be placed into sacks.

A Tarahumara miner removing geodes from a particularly thick geode-producing zone.

Deep underground in the Chihuahuan Desert, Tarahumara Indians are equipped with a pick, shovel, and wheelbarrow. The pick is used to work the walls of the pay zone, removing the geodes. The shovel is utilized to remove the waste rock from below the working wall and the wheelbarrow is used to haul the waste rock to abandoned portions of the mine. The geodes tend to occur in "pods" and once the first geode is found, several others are likely to occur with it. It's analagous to picking potatoes. The pods occur sporadically, and once a pod of geodes is removed, it may be one foot or three feet before another cluster is found. In other words, prediction is impossible.

It is estimated that one-half to three-fourths of the geode-producing horizon, and contained geodes, must remain in the mined area to provide structural support. Active mines utilize either a metal or wood head-frame with a rope and bucket attached to remove the geodes and underground country rock. This mechanical set up is referred to in Spanish as *La Cigueña* – the stork – an appropriate name for

Waste rock or "muck" is removed from below ground through large drums such as this.

The geodes are placed in sacks underground and then hoisted to the surface. Each sack weighs between 100 and 150 pounds.

this delivery system. The process is similar to how water is drawn from a hand-dug water well. The clay-like ash flow tuff, which contains the geodes, is removed through the use of picks. The waste material which is not used for back-fill is then transferred up the mineshaft and is disposed of as mine dump (mine tailings). Over 400 buckets (each weighing 100 to 150 pounds) of waste rock are removed daily from the underground workings.

Currently, miners produce approximately ten tons of coconuts per week from an active mine. However, only two out of ten coconuts produced will be hollow. Miners are paid a daily flat rate and must produce at least two sacks of coconuts per day. Each sack of coconuts weighs 130 to 150 pounds. A bonus is paid for each sack above the required two. The workweek is six days. During the geode-mining heyday of the late 1960s and early 1970s, upwards of 60 to 70 miners and family members lived and worked at Las Choyas. Today that population is around 20. There were also upwards of 20 different active claims, all owned by Hector Carrillo, Tony Krahmer, Victor Salgado, Gerald Berry, Rodolfo Castro, Tacho Solis, Gabe Olvera, or Chuy Pacheco.

Below the surface, coconuts are placed into plastic sacks, each bag containing 150 pounds of material. The bags are then hoisted to the surface by steel cable. The geodes are then loaded onto a flatbed truck and taken 18 miles away to the Carrillo ranch house (Rancho El Sacrificio) for sorting and processing. It is at the ranch where the crucial determination is made as to which coconuts are hollow and which are solid.

The miners at Las Choyas are absolutely superb at both the mining and sorting process. Based upon experience, they can tell you with a high degree of certainty what size the hole will be in a geode based upon hefting the nodule. However, no one, not even the miners, can tell you which ones will contain amethyst and which ones won't. Rough geodes are not only sold as solid, semi-hollow, or hollow, but are also sold based upon the size of the geode. Geodes are sorted into three categories based upon size: 1.5 – 2.0 inches, 2.75 – 4.5 inches, and 4.5 – 6 inches. Many of the geodes are "solids" and are used for ornamental objects such as sliced and polished bookends. The price of the coconut will vary according to size and whether it is hollow or not.

The Hector Carrillo family, owners of the coconut claims, have designed and built a "sieve" or "sorting ramp" that dramatically speeds up the essential process of separating the coconuts into different and specific sizes. The ramp-like device is approximately 25 feet long and 3 feet wide and is primarily made up of 3/4-inch diameter steel rods. It looks like a "ramp" in that it's sloped from a height of about five feet on one end to one foot on the other. The surface is separated into five-foot sections of parallel steel bars running along the length of the entire device. The bars of each individual section are spaced in such a way as to provide openings of 1", 1.5", 2.5", 3.5", and 4.5" with the smallest set of openings at the top and each subsequent section of openings getting progressively larger.

The higher end is where the sorting process begins when the bags of rough coconuts, of all different sizes, are poured onto the top of the ramp. From there, gravity, an occasional persuading shovel, and very quick hands take over, all encouraging the almost perfectly spherical geodes to roll down the sloped ramp. The smallest geodes are the first to fall through the highest section of the ramp. Then the 2.5 inchers, the 3.5 inchers, and so on until the end of the ramp where geodes too big to fall through any of the openings roll off the end into their own pile. Beneath the ramp are plywood dividers which are aligned with each size section. These dividers keep the sorted geodes contained in their own size group

Once at the surface, the mine-run coconut geodes are sorted according to size and percentage of hollowness. This "sorting ramp" dramatically speeds up the process of separating the coconuts into specific sizes.

After the geodes are sorted according to size, each one is 'hefted' by hand to determine whether it is hollow or not.

and ready for the next step, which is distinguishing the hollows from the semi-hollows and solids. Each geode is hefted by experienced helpers to determine its hollowness.

It is easier to determine relative density of hollowness by comparing two geodes of about the same diameter. To accomplish this, a geode is placed in each hand, bouncing each geode and comparing the relative weight. Geodes of the same size and density are then bagged into 100-pound lots for shipment to El Paso where they are wholesaled to destinations throughout the world. Because of the exceptionally large amount of "solids" coupled with a low demand, most of the solid nodules remain at the Carrillo ranch.

The land, like any other Mexican agate or mineral deposit, has claims filed upon it. Access to the geode locality is very difficult, at times impossible. The

These 'mountains' of geodes at the Carrillo ranch are discarded 'solids' that have been through the sorting and hefting process.

The rough hollow geodes are packed into drums for shipment to worldwide destinations.

roads leading into the area are poor, and during the summer months they are usually washed out due to heavy afternoon thunderstorms. A second obstacle also exists: strongly locked gates.

The Las Choyas geodes were first discovered in 1960 by a cowboy (vaquero) who noted perfectly spherical rocks lying on the ground. The land, Rancho Mesteño, was owned by Trinidad and Simona Carrillo. In 1961, Sr. Ramon Peña of Cd. Juarez, Chihuahua, filed the first claim in the area; however, it was not until 1965 that the first production of geodes occurred. It was Jim and Joe Miles of El Paso, Texas, who sold Ramon Peña's first lot of geodes from Las Choyas. The geodes were offered for sale at the 1965 Phoenix Gem and Mineral Show. Initially, the geodes retailed for five to ten cents per pound for solids; hollows sold for 25 cents per pound. (Miles' initial sorting was based on rolling the nodules on his concrete patio and listening for a difference in sound. The hollow geodes made a slightly different noise than did those that were solid.) Today's prices are based upon the size of the geode and its percentage of hollowness.

Just after this historic 1965 event, claims sprang up all over the two-square-mile area. Through time, those half dozen or so claim owners have dwindled to two. Hector and Jeannette Carrillo of Gem Center U.S.A. are not only the landowners, but operate the most productive mines in the region. Jeannette is the daughter of Ed Barry, a pioneer gem and mineral dealer in the El Paso area. As a teenager, Jeannette would travel with her dad to the remote area of Las Choyas and purchase rough geodes at the mine. It was there that she met Hector, the son of ranch owners Trinidad and Simona Carrillo. One of Hector's primary responsibilities at that time was sorting geodes. A romance developed and the rest is history. Today, their sons continue in the same tradition, making the coconuts a multi-generational operation.

Las Choyas coconuts go by many names including baby coconuts, cocos, hollows, solids, coconut geodes, Chihuahua geodes, and Mexican coconuts. Especially nice geodes are many times referred to as "old originals," and exceptionally large ones are sometimes advertised as "jumbos." Wherever you go in the world and see a coconut geode under any given name, you will know it came directly from the one and only Carrillo ranch.

Opening "Coconuts"

It has been said that opening coconuts is like putting coins into a slot machine. Will this one be a winner? Mexican coconuts can be sawed with a diamond saw or opened with a pipe cutter made for cutting cast iron sewer pipe. The pipe cutter allows for tightening the grip on the geode prior to applying pressure much like the screw on the handle of vise grip pliers. One end of the chain is attached and the other is loose but fits into a hooking-type device to hold it. The chain is tightened by hand over the middle of the geode, then using an adjustment screw, the chain is tightened a bit more. The handles on the cutting tool are about three to four feet long. One handle is placed on the ground and is stood on while pressing down on the other until the geode cracks. This technique was perfected by the late Dalton Prince of Houston, Texas. Opening small geodes in this manner makes a wonderful event at children's birthday parties. Not only does it provide great party entertainment, but party favors like this are hard to compete with.

In sawing geodes, first look for the largest dome on the specimen. It is believed this dome is in the upright position when the specimen is forming. Saw

The late Dalton Prince of Houston opens geodes with a pipe cutter made for cutting cast iron sewer pipe.

The audience marvels at the beauty, knowing they are the very first to have a glimpse of the hidden treasures inside.

through this largest dome, and it is likely you will expose the best "picture" or surface. If the specimen is elongated or egg-shaped, saw lengthwise in order to obtain the best exposure.

After being cut with a lapidary saw, Mexican coconuts are usually placed on a vibrolap automatic lapping machine. An electric motor drives an eccentric shaft connected to a flat lap plate, causing it to vibrate or oscillate so rapidly that geodes placed on the surface remain almost still while the lap moves beneath them.

The author's son uses a lapidary saw to open this coconut geode.

After being cut with a lapidary saw, the geodes are placed on a vibrolap automatic lapping machine.

Coarse, loose abrasive grit is initially used in order to level the geode surface as quickly as possible and then to finish with grit fine enough to permit polishing afterward. The time required for coarse lapping is from six to seven hours and from five to six for fine lapping. Polishing is done on the same machine with a soft polishing pad installed and charged with polishing powder and water. Polishing takes an additional three to four hours. Although many hours are consumed in completing specimens from lapping to polishing, very little is hand labor, and the large area available permits treating a number of specimens at the same time.

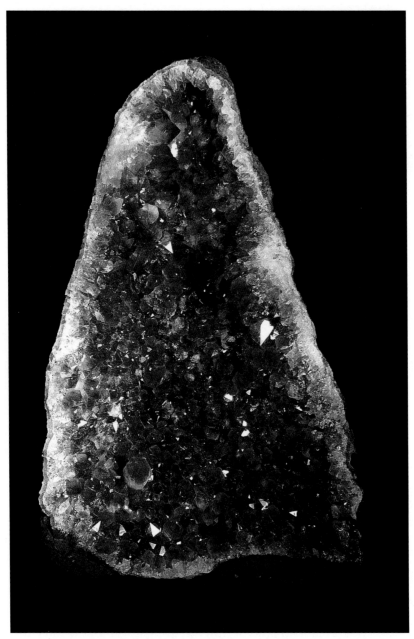

Amethyst cathedral from Rio Grande do Sul, Brazil.
(Photo by Donnette Wagner; Specimen Courtesy of Donnette Wagner)

CHAPTER FOUR

When Geodes are Gems

— BRAD L. CROSS —

BRAZILIAN AMETHYST GEODES

Rio Grande do Sul is the southern-most state of Brazil and is found along the sunny Atlantic coast. It borders the state of Santa Catarina on the north, Uruguay on the south, and Argentina on the west. It is here in this southern state that some of the world's most extensive amethyst and agate deposits are found. The amethyst geode deposits are found primarily in the northern half and western portions of the state, south of Iraí in the Ametista do Sul (São Gabriel) area in a very well defined basalt horizon at 1,300 to 1,450 feet above sea level.

Geode mining in this area dates back to the mid-1800s when European immigrants discovered and began to work the region's deposits. They shipped tons of agate and amethyst geodes back to Germany in cattle hides as ballast aboard sailing ships. As word of the strikes spread, other deposits were discovered throughout the region. Much of this area was and still is nearly impenetrable due to dense forests, and roads, where they existed, were usually quagmires that could not support heavy wagons. As a consequence, the accessible areas were mined first and even today new deposits of considerable size are being found in Brazil and Uruguay.

The town of Soledade is centrally located in the heart of Brazil's amethyst and agate country. It is the cutting center for the region's gems, where huge warehouses contain forests of tall, short, and wide geodes that tantalize even the most sophisticated viewer. More than 125 million years ago this region, with adjoining areas of Uruguay, Paraguay, and Argentina, experienced extensive volcanic activities that produced one of the largest discharges of lava on the surface of the earth.

43

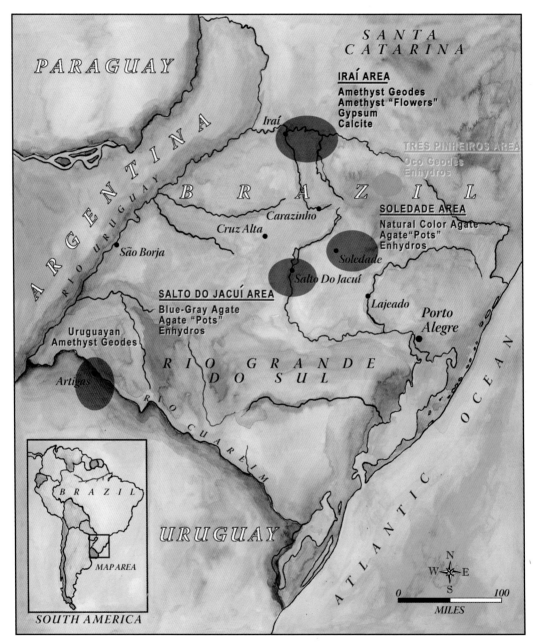

Amethyst and Agate Geode Occurrences in Rio Grande do Sul, Brazil.
(MAP CBY MARK LESH)

The geode-producing geologic formations of Rio Grande do Sul are those of the Paraná Flood Basalt Province. The Paraná basin is the dominant geological element of southern Brazil and extends from São Paulo through Paraná, Santa Catarina, and Rio Grande do Sul to Uruguay and Paraguay.

Amethyst and agate form in vesicles within the lava flows of the Paraná Basalt. (The flood volcanism extends over an area of 745,644 square miles.) This extensive lava series has an average regional thickness of 2,100 feet but reaches a maximum thickness of 6,000 feet. It isn't unusual to see an individual flow up to 300 feet thick and covering a distance of several hundred miles.

The Paraná basalts erupted in an extremely arid desert environment that was dominated by sand dunes as evidenced by the extensive red sandstones (the Botucatú Formation) found at the base of the volcanic sequence. There were also quiet periods during the various basalt floods during which layers of sand and sediment built up, separating the various flows. The fine-grained basalts which host the geodes are composed of the minerals plagioclase, pyroxene, magnetite, and volcanic glass.

The Paraná basalts formed in conjunction with the opening of the South Atlantic. The Jurassic to Cretaceous (138 to 127 million years ago) basalts ascended in fissures through the earth's crust and spread laterally over an extensive region. The old idea that these flows went whooshing over the countryside at incredible velocities (like a flash flood) have been replaced with the concept that the units are emplaced more like flows, namely slow moving and filling preexisting valleys and depressions.

Up to nine individual basaltic flows are found within the geode-producing area. However, the amethyst geodes are only located in the fifth flow, named the São Gabriel Flow (the miners refer to this unit as "basalto portador"), in a well-defined horizon (1,300 to 1,450 feet above sea level). This geode-bearing flow can be subdivided into three specific units from the top to the base:

1. A 15- to 30-feet thick weathered vesicular basalt horizon, many times containing angular sandstone fragments;

2. A 1- to 5-feet thick transition zone of highly fractured basalt; and,

3. A 65- to 100-feet thick massive basalt which can be further subdivided from top to bottom into a 10-feet thick layer which contains the geodes; a 30- to 50-feet thick massive basalt; and a 10- to 15-feet thick highly fractured basal unit.

The amount and the quality of the amethyst from this basalt unit have turned this area into the main supplier of amethyst to the international market. Although production records are not kept, it has been estimated that approximately 100 to 200 tons of geodes per month and 20 to 30 tons of amethyst crystals for cutting per month are produced in the region. The geodes from the state of Rio Grande do Sul in southern Brazil are usually elongated vertically in a chimney, cylindrical, or pipe-like form. Most have a spherical cap-like head shape towards the top. The base of the geode is generally dimpled, upward convex. The pointed spherical cap-like shapes are always found aligned in an upward position in the basalt flow. Due to their unusual shape, some folks refer to these geodes as "cathedrals" and "amethyst tunnels." Once an expensive rarity, cathedrals are now a common sight at shows and shops catering to gem and mineral enthusiasts.

The formation of these amethyst geodes followed a two-stage process that is very similar to the Mexican coconuts. The early magmatic stage is responsible for the generation of the cavity with the second phase following millions of years later with a very low temperature mineralization or cavity filling.

Lets look at the first stage, the magmatic formation of a cavity. There is no question these are unusual shaped geodes. How do you get a shape like this; and why are they all aligned in a specific horizon within the upper part of the basalt flow at the same elevation throughout the region? Numerous studies have shown that shapes and features such as this can be explained by the ascent of a fluid with a lower density and viscosity than the surrounding medium as well as an increasing viscosity from the bottom of the flow to the surface.

Volatile constituents are dissolved in every magma. These are mainly carbon dioxide and water. When the magma rose through the fissures in southern Brazil the solubility of the volatile components in the magma decreased. The volatiles were exsolved and gas bubbles formed. Exsolving gas bubbles can coalesce quite easily in a magma with low melt viscosities. Upon further ascent in the flow, fluid bodies can eventually split into smaller cap-shaped bodies of similar size – shapes exactly like we see in the Brazilian geodes. Once referred to as "spike amygdules", these Brazilian vesicles can be man-sized and it's difficult to explain their size by the formation of gas bubbles from a basaltic magma alone. Some geologic models indicate that the hot lava reacts with cool water from rain, rivers, or from underlying aquifers which may have circulated through the cooling lava flows.

The second step in the formation of amethyst geodes is the filling of the cavities. Several recent studies utilizing Potassium-Argon isotopes to determine the age of the outside layer of celadonite, thus providing us with clues on the age of the geodes, suggest that the final filling of the large amethyst geodes occurred some 40 to 60 million years after the formation of the cavity (Vasconcelos, 1998). These same studies indicate that geode infill is a long-lasting process, on the order of millions of years. It has also been demonstrated that the fluids which were present during the filling of the amethyst geodes were very low temperature, less than 200° F.

A good question to ask is, "Where do the geode-forming fluids come from?" The late timing of the actual geode formation with respect to the formation of the cavity strongly suggests a meteoric source for the mineralizing solutions. However, it's important to remember the extensive red sandstone underlying the volcanic sequence. This sandstone is one of the world's largest freshwater reservoirs, the Guaraní aquifer system. The water temperature of this artesian aquifer is comparable to the temperature of formation for the mineralizing solutions and should not be excluded as a possible source for the solution filling of the amethyst geodes.

Now that we have a fluid to infill the cavity, what is the source of silica to form the agate rim and internal amethyst crystals? The most likely silica source is the highly reactive interstitial glass in the host basalt. Additional components in the basalt are an excellent source for the calcite and other secondary minerals sometimes found in the geodes.

Without question, some of the world's largest and most beautiful geodes are found in these South American countries. History's greatest amethyst geode was discovered in 1900 in the Serra do Mar, about 87 miles north of Santa Cruz in Rio Grande do Sul. The geode was far too large to remove intact – 32-feet long, 6.5-feet wide, 3-feet high, it weighed about 35 tons. Ten sections weighing a total of 15 tons were shipped to Idar Oberstein, West Germany, reassembled, and shown in the Dusseldorf Exhibition of 1902 where deep violet-colored crystals, averaging 1.5 inches in diameter, caused a sensation. This region of Brazil has produced at least one hundred times more amethyst than the next most prolific locality in the world. It is so abundant that almost every private collector and museum has at least one specimen.

An added bonus is found some 30 miles south of the Ametista do Sul in the Salto do Jacuí area. There are other Brazilian geodes, including agate geodes. The Salto do Jacuí area is host to the world's largest deposits of agate. While they may not be the most colorful ever found, there is no site anywhere which rivals the quantity and size of these banded agate nodules. Some of these agate nodules aren't quite completely filled with agate and have a small pocket lined with sparkly, drusy quartz crystals in the center. It's important to note though that while the monstrous amethyst geodes and the Salto do Jacuí agate geodes are neighbors, they are never found in association with each other. They are separate and distinct occurrences.

Visiting Brazil's amethyst and agate mines means an often arduous, though colorful, trip. I flew into São Paulo, Brazil where I connected with a flight south to Porto Alegre, the capital of the state of Rio Grande do Sul. The drive to Soledade was some three hours long, cutting through some of the most beautiful country to be found anywhere. The landscape shifts from coastal lowlands to a horizon dominated by subtropical-forested hills. A territory of watercourses winding through diverse vegetation provides newcomers with a whole new set of cultures of unimaginable exquisiteness. A deep red soil is found here that hosts some of the most beautifully wooded areas one can ever imagine. Exotic ferns and fancy colorful mushrooms abound. There is more—animals, unique beasts such as tapirs, capybara, pacas, and sakis, and wonderful birds – flocks of parrots, jaburus, and toucans. A source of pleasure for any taste!

Here there are vast cattle ranches, sources of the succulent beef that has made the gaucho barbecue – the churrasco – a synonym for peppered steaks and sausages throughout Rio Grande do Sul. It is always served with cabbage, rice, and slices of onion—no menus or special orders here.

Mineralogy

Most amethyst geodes from the Ametista do Sul region display the following mineral sequence: (1) a very thin dark green exterior of fine-grained celadonite, which facilitates their extraction during mining; (2) a less than one-inch wide microcrystalline massive quartz or banded agate layer; (3) a less than one-inch thick layer of incompletely crystallized and fractured colorless to milky quartz; and, (4) a layer of amethyst showing progressive color zoning which continues from the last colorless quartz layer to the darker purple quartz. Within each individual crystal, bands of amethyst color many times alternate with colorless bands having light and dark patches. The color can be modified, giving a fine orange color. These heat-treated amethyst geodes are often sold as citrine geodes.

Late stage or secondary minerals are represented by calcite, selenite, and barite that occur on top of the silica minerals. The youngest generation of calcite in the geodes display a variety of crystal habits ranging from flat rhombohedra to elongated scalenohedra and long prismatic crystals. What an eye catcher to find a bright purple amethyst geode with sprays of perfectly terminated dainty white calcite! Some of the geodes contain calcite crystals and occasionally quartz pseudomorphs after anhydrite crystals up to six inches long. The pseudomorphs are rectangular and taper to thin, flat tips.

Both Brazilian and Uruguayan amethyst geodes commonly contain inclusions of radiating crystal groups or tufts of long, thin crystals that range from needlelike to fibrous. These golden yellow goethite crystal groups stand out in relatively high relief within the purple quartz. The long yellowish goethite crystals outline an earlier stage of development of the amethyst crystal. This indicates that the goethite crystals were growing upon the faces of the amethyst crystal while it was developing, and it incorporated the goethite in the process.

Color zoning is also common in the Brazilian geodes. Zoning is observed as alternating shades of purple and colorless layers (zones) that are arranged concentrically parallel to the external crystal faces and are caused by successive stages of growth.

Oxygen isotope studies (Juchem, 1999) performed on agate, colorless quartz, and amethyst have not revealed any significant variations among the mineral phases analyzed. As a result, the silica mineral phases might have crystallized from an original fluid that was characterized by a constant temperature under stable geologic conditions.

Mining

The heaviest concentration of amethyst is in the Ametista do Sul region within a 186 square mile area and includes as many as 500 mines or diggings (*garimpos*). The most active mining areas are found near the municipalities of Ametista do Sul, Planalto, Iraí, Frederico Westphalen, Alpestre, and Rodeio Bonito, where the properties are primarily mined by larger firms such as Jaghetti, Willi Guerner, Ledur, and Irmãos Lodi. Within these regions, it's not uncommon to find lichen-encrusted agate nodules and broken geodes with drusy quartz lying in the soil.

Local miners (*garimpeiros*) prospect for amethyst in open pits as well as underground tunnels or galleries, some up to 150 feet deep. The tunnels are driven parallel to each other into the cavity-rich basalt flow that has been exposed along the hillside. The miner's drift, that is, dig horizontally, with explosives, hydraulics, and a lot of very hard labor. The tunnels are usually interconnected underground with many cross cuts. The mines seem like endless underground rooms with a large number of big arching pillars. The varying size and frequency

In the initial phases of mining Brazilian amethyst geodes, the hillside is terraced to allow easier access.

A series of tunnels or galleries are driven parallel to each other into the cavity-rich basalt flow that has been exposed in terracing operations. The tunnels are usually interconnected underground with many cross cuts. The mines seem like endless underground rooms with many big arching pillars.

of the amethyst pockets, as well as the quality and size of the individual amethyst crystals, is truly baffling. However, like all other geode deposits, not all of the specimens encountered are hollow. Many are simply solid quartz.

Miners drill holes in the dense basalt, pack them with black powder or explosives, and slowly remove the host rock. Utilizing a pneumatic hammer, additional basalt is removed until the hammer suddenly lurches forward, indicating the point has broken into a hollow amethyst geode. The miners will now proceed carefully if they wish to remove the geode intact. Enough rock must now be chiseled away so that the miners can look through the small drill hole to inspect the quality of amethyst.

Inspection of the geode is accomplished by attaching a car battery to a small automotive light bulb located at the end of a three- to four-foot electric wire. Through a small half-inch hole, the cord and light bulb are slowly pushed inside

Once a geode is discovered, it is removed through the use of a hammer and chisel.

A small hole is chiseled into the geode and the specimen is inspected by attaching a car battery to a small automotive light bulb at the end of an electric wire. Peeking inside the small hole, the miner is able to determine the quality of amethyst and the viability of removing the geode.

the geode to illuminate its sparkling interior. Peeking inside the small wound, the miner is the first to view one of the most spectacular sites ever imagined — an isolated void, sometimes the size of a small room, lined with intense and perfect purple crystals. With the interior lighted, the quality of the amethyst and the size of the geode is determined. If the amethyst is pale and of poor quality, the miners often ignore it and continue to drill and blast without any attempt to save it. If the amethyst is of high quality, the miner will laboriously chisel away, trying to remove the geode intact. This involves anywhere from several days up to a week of handwork with hammer and chisel — lots of muscle and lots of patience.

The Brazilian geodes outer surfaces are coated with dark green micaceous minerals such as celadonite, and they break cleanly and fairly easily away from the encompassing basalt. Most geodes from Uruguay have little to no outer mineral coating, making them extremely difficult to separate when found encased in basalt.

Geodes which contain large quantities of gem grade amethyst are many times broken into sections and removed from the surrounding rock matrix as quickly as possible. They are then taken to a secure area where individual crystals will later be transformed into facet-grade rough. A pocket of gem grade amethyst worth several thousands of dollars can be an irresistible overnight temptation for a miner whose salary is only several hundred dollars per month.

The geodes are weighed in the field on crude scales and the miners are paid based upon the production and quality of the geode. The specimens are then taken to cutting factories in Soledade, Lajeado, or Estrela where they are either cut open with a diamond saw or tapped open with small specially designed picks. The opened geodes are then either packed into fifty-five gallon drums or crated in special wood boxes. Sawdust is carefully packed around the geodes for shipment to worldwide destinations.

At the cutting centers, those geodes that have crystals with good color and high clarity are often broken into plates. The plates are further broken down into individual crystals. These single crystals will soon become flawless facet grade rough. This is the material which sells by the gram, not the pound. Individual amethyst crystals are held between the fingers and with a small special hammer, known as a swing hammer, the imperfections within the crystal are cobbed away. Experience is the key to successfully cobbing amethyst. It is not at all uncommon to see a man in a Soledade or Lajeado cutting factory with 25 years experience in cobbing, work a single crystal down to a small pea-size flawless stone within a matter of minutes. From here, cutters fashion the rough into a sparkling stone to be worn in a lovely ring, a bright pendant, or a brilliant brooch—all from the inside of a sparkling geode.

Opening Brazilian Geodes

Like the Mexican "coconuts," Brazilian geodes may be sawed open (usually requiring a 36-inch diamond saw) or they may be "scored" open. Because of their irregular shape, a pipe cutter isn't an option as it is with coconuts. Instead, small pointed picks are utilized to "tap" or "score" open the geode.

The Brazilian geodes are first marked down the middle with a chalk line so that two equal halves are defined. Utilizing a small pointed pick or hammer (similar to a small welders pick), the miners gently tap along the chalk line scoring the geodes with the sharp pointed tool. As small fractures develop perpendicular to the scored line, an epoxy-type glue is used to penetrate and seal the fracture and keep the geode from splitting into numerous plates. The chalk line is carefully followed until the entire circumference is deeply scored with the pick. The geode is then carefully pried apart with a large flat-head screwdriver revealing the rich inner beauty.

Those specimens which are cut are usually done so using a 36-inch lapidary saw. Once the geode is cut down the middle, the specimen is sanded and polished by hand. The smaller beauties are taken to large horizontal belt sanders where the specimens freshly cut surface is sanded on 220- and 600-grit sanding cloths. A final polish is provided through the use of a hard felt wheel and cerium oxide. The larger of the geodes that cannot be easily handled are treated using hand-held sanding and polishing equipment.

Heat Treatment

It's not at all unusual to find citrine geodes on today's market. Unless you are told otherwise, you can safely assume that it is heat-treated amethyst. The question of color in crystallized quartz is a fascinating one, the lovely colors being related to the internal atomic structure and its imperfections. Citrine and amethyst also owe their color in part to background radiation. The radiation primarily occurs due to naturally occurring Potassium-40 and members of the uranium and thorium

Some geodes are opened with the use of a special pick hammer. Small fractures develop during this process and must be sealed. Opening this geode will take anywhere from 30 to 45 minutes.

Once opened, the inner beauty of the geode is revealed.

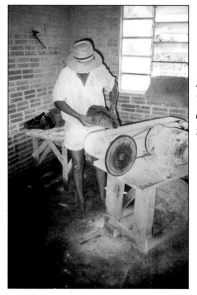

Some geodes are opened with the use of a large lapidary saw. The specimen is then sanded on a belt sander. The belt sander consists of an endless belt of abrasive cloth passed over two rubber-faced pulleys.

After a thorough sanding, the geode is polished on leather with cerium oxide.

Some geodes are so large that the sanding and polishing equipment must be brought to them.

The polished geode is then thoroughly scrubbed with soap and water to remove sanding grit and polishing compound.

The geodes are finally packed with sawdust into 55-gallon drums for shipment to world-wide destinations.

decay series found in rocks and soil. As is true for amethyst, the formation of citrine requires the presence of trace amounts of iron impurities in the quartz. The difference between citrine and amethyst is the oxidation state of the iron: $Fe+3$ in citrine and $Fe+4$ in amethyst.

Subjecting amethyst to heat will reduce the oxidation state of the iron impurities in the quartz structure thereby causing amethyst's purple color to fade and become yellow to reddish-orange, green, or colorless depending on the amount and duration of the heat applied. The purple color of amethyst usually can be regained by irradiation which reoxidizes the iron impurities. Nature produces the same shades of amethyst and citrine that heat-enhancement creates, and in a finished stone it can be difficult to determine whether the color is natural or heat-induced. One telltale sign is that inclusions may expand under controlled heat, causing minute fractures in the stone.

While at the cutting factories in Brazil, I found various-sized kilns, like the furnaces used by a ceramics worker, to heat the amethyst to citrine. Some of the kilns were small and were used for treating facet-grade amethyst. Others were large and could hold a nice sized geode. The thermostats on these ovens were adjusted to bring the stone slowly up to the desired temperature, then return the temperature slowly back to normal.

The results were astonishing, all dependent upon the amount and duration of heating. Some very dark amethyst was simply reduced to a lighter shade, while

Many high quality amethyst geodes are cobbed into flawless facet grade rough. These pieces will ultimately be faceted and used in jewelry. Two small ovens are shown here which are used to heat treat the amethyst into citrine.

the very light amethyst was heated to produce citrine and, to a lesser extent, green quartz. If the amethyst contained color zoning, even more dramatic color changes were observed. Because the results of heat treatment on these geodes are considered stable and permanent under most conditions, no special care is needed.

Flowers

A discussion of Brazilian geodes would be incomplete without mention of amethyst flowers. These little treasures are primarily found in the Getúlio Vargas region of Rio Grande do Sul and are classic flower-like formations of radiating medium to light purple amethyst crystals arranged in a somewhat circular pattern. At times there will be multiple "blooms" closely resembling clusters of flowers. These are literally the "flowers of the mineral kingdom."

Warehouses in southern Brazil contain thousands of amethyst geode specimens.

Valuing

The price per pound of these geodes depends mostly on color, clarity, and brilliance of the crystals. The darker color usually means a higher price per pound. There are other factors as well, such as thickness of the wall—the thicker the wall, the less cost per pound. Is the geode cut and polished? Polishing the geode will increase the cost per pound.

URUGUAYAN AMETHYST GEODES

The sky is blue with white fluffy clouds that cast scattered shadows on the green, gently rolling plains. Meandering streams break the pampas grassland and dark green stands of eucalyptus provide shelter for cattle and sheep. This is ranching and agricultural country which appears European in many ways with many of its original immigrants from Germany, Spain and Italy.

The host rock for the Uruguayan geodes is the same as that of Brazil, the Paraná flood basalt. However, the geode-producing flow known as the Serra Geral Formation in Brazil is known as the Arapey Formation in Uruguay. The Uruguayan amethyst deposits are located at Artigas in the northwestern part of the state near the border of Brazil.

Like the Brazilian amethyst geode deposits, commercial mining is conducted where the greatest concentrations of amethyst are found. Most of the mines are concentrated in a 62-square mile area about 19 miles southeast of Artigas. Uruguayan amethyst has always enjoyed the reputation of being darker than Brazilian amethyst. However, in reality, much Uruguayan material is not of good color.

Mining

Unlike the Brazilian deposits, few Uruguayan geodes are removed intact. Most amethyst in Uruguay is dug from the soil. The amethyst, like that in Brazil, originally formed on the walls of vesicles that had been trapped in the basalt when it cooled. After millions of years of weathering, the upper portion of the basalts have turned to soil (in some places 30 to 40 feet thick) and contain scattered concentrations of amethyst. A backhoe is usually used to dig holes approximately 15 feet deep, stacking the dirt to one side.

As the backhoe digs into the dark sticky clay, an occasional clunking or grating noise is heard and frequently a white scuff mark is left on the earthen wall. Miners carefully check the exposed surface to determine if it is amethyst or simply partially decomposed basalt. Pointed steel bars are used to pry amethyst plates from the working face. Due to extensive weathering, most of the amethyst geodes have collapsed but are still together as groups or nests of amethyst plates. The plates are covered with thick and sticky mud and are taken outside the pit where they undergo a series of wash and dry cycles. The amethyst is not left exposed to the sun for more than a few days because its color will eventually slightly fade.

The plates have been buried in the soil for thousands of years and are often stained reddish-brown with iron minerals. To clean the specimens, local factories use a Brazilian industrial cleaning solution called *Chispas*, which they buy in 55-gallon drums. This solution contains a small amount of hydrofluoric acid so the length of time that the amethyst specimens remain in the solution is critical. The agate skin on the Uruguayan amethyst is much more susceptible to attack from the acid than is the amethyst and the solution will many times leave a white "skin" covering any portion of the agate it comes in contact with. Few complete Uruguayan amethyst geodes have survived after thousands of years of weathering. The few geodes that do survive may show cracks when they are cut open.

Although most amethyst from Uruguay is dug from the soil, several hard-rock mining operations do exist. The mining operations and procedures are identical to those of their cousins in Brazil. However, most of the Uruguayan geodes do not have the outer coating of dark green celadonite and they cling tenaciously to the basalt. To remove the geodes intact, a generous layer of basalt must be removed with the geodes to ensure that they don't break.

"Stalactites" and "Bulbs"

Some of the most striking Uruguayan specimens are the amethyst "stalactite" sections that frequently have a core of concentrically banded agate, calcite, or quartz with a 360° rim of amethyst crystals. These amethyst "stalactites" are formed when amethyst crystals grow as a coating over agate or calcite tubes that

Amethyst stalactites and perimorphs are sometimes found in Brazilian and Uruguayan geodes.

(PHOTO BY KIM JONES; BRAD L. CROSS COLLECTION)

have occasionally grown inside the geode during formation. Many of the damaged or lightly colored stalactites, not suitable for specimen display, are sliced some 3 to 4 mm thick, are polished on both sides, and are used for exquisite jewelry. The variety of size and color combinations is amazing, each cutting sections that are different from those cut from any other.

High quality amethyst stalactites are rare. Only one in ten are free from damage, and of these, only one in ten will have good color. Only one in ten of these select few are lustrous and shiny. This all boils down to about only one in 1,000 being a mineral collector's fantasy.

Sometimes the specimens are no more than little "bulbs," "knobs," or "humps" of amethyst and some even have a hollow pointed cavity where a scalenohedral calcite crystal used to be (a perimorph). It's also not uncommon to find hollow hexagonal columns of drusy amethyst crystals with flat tops. Some casts still have the calcite in them. What an incredible mineralogical world!

CHAPTER FIVE

More Latin Surprises

— BRAD L. CROSS —

exico and Brazil are vast nations and many of their natural resources await discovery. While maybe not as large as the more well-known deposits, smaller occurrences of unique geodes do exist. Several of these tantalizing Latin deposits are discussed in this chapter.

TABASCO GEODES

Also called "ilianites," these geodes were once a hot jewelry item. Popular in the mid- to late-1970s, these geodes were cut in half, polished, and the rind was coated with 24K gold electroplate. They were fashioned into pendants and earrings of all colors.

Tabasco geodes were first introduced around 1975 in the city of Zacatecas, Zac., Mexico. The owner of Rancho Agua Blanca near Tabasco, Zacatecas (some 80 miles south southwest of the city of Zacatecas) was the first to actually mine the geodes from his ranch. Due to poor marketing, production stalled until Luis Arzola of Cd. Juárez, Chihuahua, brought the sparkling gems into the United States, naming them after his daughter, Iliana. Thus the name, "Ilianites." The late Delma Perry, a popular mineral dealer in El Paso, was the first to form and electroplate the polished geodes into jewelry pieces.

Due to their immense jewelry popularity, production increased drastically. The demand and popularity caused a serious rift between the rancher, his son, and the various mineros (miners) who were digging at the ranch. Production ended as quickly as it began, all due to greed.

61

Tabasco geodes from Zacatecas, Mexico.
(Photo by Donnette Wagner; Brad L. Cross Collection)

Tabasco geodes' extremely smooth, dark to bright green exterior makes them easily distinguishable from any other geodes and nodules. Tabascos are small, averaging less than one and one-half inches although a few reach upwards of two to three inches. Some are as tiny as two-tenths of an inch. All of these little gems are very thin skinned, with their rinds usually averaging under one-eighth of an inch thick. The interiors are extremely colorful, ranging from sky blue to rust to black. A bright, sparkling drusy quartz coating lines each one.

The host for these lovely little gems is a vesicular basalt. Weathering has progressed to the point of where the basalt is slightly weathered, but far from the point of a clay, like that found at the Las Choyas coconut beds.

The geode-producing basalt ranges from six to twelve feet in thickness. Few geodes are found in the top portion of the volcanic unit, but are instead heavily concentrated in the basal three to four feet of the formation. What basalt vesicles that are at or near the top of the unit are generally filled completely, producing only solid nodules.

Ninety percent of the geodes that range in size from one-half inch on up to one and one-quarter inch are hollow. This size is the most ideal for jewelry purposes. The smaller geodes as well as those over one and one-half inches are mostly solid.

A backhoe is used in the mining process to dig down to the bottom of the unit where the best geodes are concentrated. Chunks of basalt are broken up by hand, freeing the geodes, which are then screened by hand. The geodes tend to fall out of the matrix although some do require a bit of work with the hammer to be freed. Unfortunately, a fair amount of the geodes are broken in this process. Reportedly, it is possible to produce as much as 200 pounds of Tabascos per week.

The deposit currently covers only an acre or so, but may possibly continue farther under a nearby hill. However, with a greater amount of overburden to remove, the deposit will no longer be economical to mine and operations will probably cease.

The first and only official claim (denuncio) was filed in 1995 by Ruben Avila Contreras (Cd. Juárez, Chihuahua) and in partnership with Mike New. The claim is named "El Papa" after the Catholic pope.

TRANCAS GEODES

It was a typical 120° summer day in 1972 when Sr. Erasmo Hernandez of Aldama, Chihuahua, Mexico was tending his cattle and found shattered pieces of sparkling rock on the ground. His ranch, Rancho Trancas, is some 80 miles south of the world-famous coconut geode deposit. Little did Sr. Hernandez realize what a geode was or that he had discovered a new geode deposit that would contain some of the most unique forms of quartz found in Mexico. His find also extended Mexican geode occurrences considerably farther south than previously thought.

Rancho Trancas, named after nearby railroad Estación Trancas, is about 5,600 acres in size and is located about 12.5 miles northeast of Aldama or 25

Twisting tree-like growths and ram's-horn projections make this Trancas geode a striking specimen. (PHOTO BY DONNETTE WAGNER; BRAD L. CROSS COLLECTION)

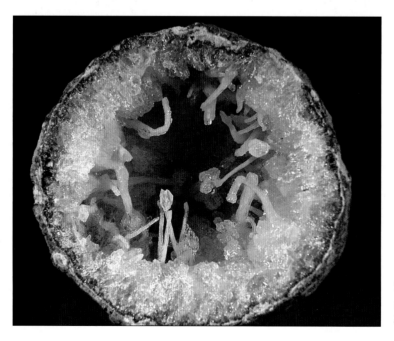

Trancas geode.
(PHOTO BY JEFF SMITH;
JEFF SMITH COLLECTION)

miles northeast of Chihuahua City. The geode deposit is claimed under the name "El Junior" and is situated just over a mile off the southeast flank of Sierra Gomez, a limestone mountain range trending northwest-southeast. Although the deposit is surrounded by limestone structures, the geodes occur in a volcanic ash flow tuff. The ash flow tuff is rhyolitic in composition and is known locally as the Mesa Formation. Potassium-argon dating places the age of the host rock at 37 million years.

Unlike the coconut geode deposit, these gems are found near the surface and can be mined through surface pit mining techniques. It's a good thing the deposit is shallow as the host volcanic rock is extremely tenacious and requires a tremendous amount of effort to remove the geodes.

When the geodes were first mined in 1973, mining amounted to using a backhoe and digging up boulders of rhyolite. The rhyolite chunks were then broken apart using hammers and chisels, freeing the spherical geodes. Once the excavated rock was worked, the backhoe was again used to unearth additional material. How ironic that these geodes were first sold to Sr. Ramon Peña of Ciudad Juárez, the same gentleman who first handled the coconuts.

Similar mining techniques were used again in 1980 when a second area was worked less than half a mile south of the original discovery. A third and final area was worked beginning in 1990 and continues to produce a wealth of specimens. The mining technique employed here involves the use of a bulldozer to cut a wide flat bench into the rolling hillside. Like the original dig, the rhyolite is ripped up in chunks where each boulder is then tackled with hammers and chisels to extract the geodes. After each successive cut is worked, another pass is made with the bulldozer to excavate more rock.

Several attempts have been made to remove geodes by driving tunnels into the host rock. By tunneling, the geode-producing zone could be mined without the effort of having to remove any overburden. This concept was abandoned not only because of the tenacious host rock, but because it was unprofitable to mine without the assistance of heavy equipment.

One of the exciting characteristics of the Trancas geodes is their fluorescence. Under short wave ultraviolet light, many of these gems fluoresce a very brilliant green due to trace uranium content. The Mesa Formation not only contains some of the most unique geodes ever found, it and the underlying Cretaceous limestone are also host to one of the world's richest uranium deposits. Apparently silica and uranium were leached from the upper portion of the Mesa Formation and migrated downward, being re-deposited in the empty amygdules

Original dig at Trancas geode deposit. Rancho Trancas, Chih., Mexico.
(PHOTO BY JEFF SMITH)

1980 dig at Trancas geode deposit.
(PHOTO BY JEFF SMITH)

found in the lower portion of the formation. The intensity of the green varies and is dependent on the amount of uranium present in the crystal structure.

As with most geodes, an array of secondary minerals can be found. Following the initial mineralization, subsequent episodes of silica deposition occurred. It is this stage of mineralization that makes the Trancas true treasures.

The secondary quartz in these geodes occurs as intergrown crystals with absolutely incredible shapes and designs. Typical hexagonal quartz crystals are rare in these specimens. Instead, rapidly changing geochemical conditions created unusual tree-like growths, curly hair-like projections, a vast array of sceptered crystals and curved formations that seem to wind in every possible direction. Tan colored opal or showy calcite can occasionally be found coating some of the unusual forms of quartz. Microminerals are also found and simply await identification by some enthusiastic researcher.

The Trancas geodes are easily recognizable by their exterior rind. Typically, they are milky white and heavily pockmarked, averaging two to four inches in diameter. When cut and polished, the thin rims display prominent white wavy or flame-like patterns reminiscent of desert roses. These Mexican gems are a beautiful and unique mineralogical wonder.

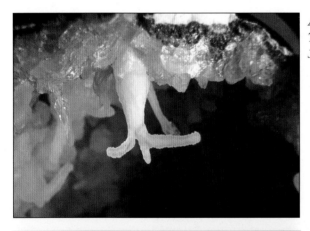

A wonderful tree-like growth in a Trancas geode. (PHOTO BY JEFF SMITH; JEFF SMITH COLLECTION)

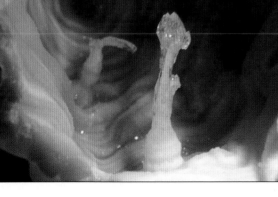

'Icy' looking, this sceptered quartz stands proudly in a Trancas geode. (PHOTO BY JEFF SMITH; JEFF SMITH COLLECTION)

SAN MARCOS GEODES

The region swallows you up in its vastness. Mile-high peaks capped with rim rock and occasional cloud shadows dwarf your body and soul. Looking off into the horizon, you see nothing but miles and miles of desert scrub. This is remote and rugged country. It's here where the ranchers pray for the miracle of rain. In the early 1980s, while working on Rancho La Prieta, a resident of Villa Ahumada, Chihuahua, Mexico found several geodes on the ground. Recognizing the potential to make a few extra pesos, the man went back home some 30 miles to the southwest and hired some friends to hand dig a 20-foot deep pit, and then a horizontal tunnel from the pit, in an effort to mine geodes. The tunnel collapsed and killed both miners. The brief mining operation ended as quickly as it began.

In 1995, Sr. Javier Villalobos Seyffert of Ciudad Juárez heard rumors of the previous mining activity and in early 1997 Javier applied for an exploration concession from the Mexican government. This document gave him access to the ranch and the opportunity to locate the previous mining operation. He found the abandoned pit and realized the potential for a significant find. In March 1997 an exploitation concession was filed for 247 acres on Rancho La Prieta under the claim name, "La Winnie," named in honor of his pet Chihuahua dog, Winnie. Over a year of bureaucracy later, the concession became official in April 1998.

San Marcos geode mining area, Rancho La Prieta, Chih., Mexico.
(PHOTO BY JEFF SMITH)

In May 1998, Javier hired men from Villa Ahumada, rented a bulldozer and a front-end loader and worked the mine for two months, producing about 10,000 pounds of mine run geodes. The geodes were introduced at the 1999 Tucson Gem and Mineral Show. The geodes are marketed under the trade name "San Marcos" and are named after Marcos Carrillo of Gem Center, USA, the primary marketer of the specimens.

The San Marcos geode deposit occurs approximately 1.5 miles off the southwestern flank of Sierra de Presidio, about 32 miles northeast of Villa Ahumada. The ground surface elevation at San Marcos is about 4,800 feet above sea level. The area is essentially flat with a slight rise in elevation from west to east.

In an unaltered state, the host rock is composed of a hard, maroon colored, densely welded ash flow tuff containing empty cavities ranging from one to four inches in diameter. The pay zone is found immediately below the tenacious welded zone in a highly altered light red to pink ash flow tuff. Alteration is extensive enough that the geodes can be plucked out with pick axes. Much like their "coconut" relatives to the south, the geode-producing zone appears to be a cooling unit, where open spaces were preserved during the de-gassing of the lava. The geodes appear restricted to the altered portion of the tuff.

After the ash flow unit was deposited and the cavities were preserved, hydrothermal solutions from nearby intrusives probably migrated upward and altered a portion of the ash flow unit providing silica as the source for primary mineralization of the geodes.

Jeffrey Smith of Pittsburgh, Pennsylvania systematically studied 187 pounds of geodes from Rancho La Prieta. The 187 pounds consisted of 235 geodes approximately two and one-half inches in diameter. Of the 235 geodes that were split open, there were 50 that were hollow or semi-hollow (20% of the nodule being hollow). These 50 specimens were examined by hand lens to select promising candidates for micromineral identification. Those selected were examined by stereomicroscope and by a scanning electron microscope equipped with an energy dispersive X-ray detector for qualitative elemental analysis.

San Marcos geodes are very similar in appearance to the Las Choyas "coconut" geodes. However, many of the San Marcos geodes tend to be slightly sub-spherical in shape. The general sequence of mineralization in most cases is a thin to thick rim of chalcedony, a somewhat thick zone of megaquartz, and finally a host of micro-minerals that appear to have formed in several stages.

San Marcos geode. San Marcos geodes tend to be slightly sub-spherical in shape.
(Photo by Jeff Smith; Jeff Smith Collection)

Minerals identified to date include: calcite (found in about 25% of the geodes as clusters of small, less than 1 mm, clear rhombohedral crystals, or as clusters of scalenohedral crystals); goethite (common occurrence as flat, bladed crystals often found in radiating or sheath-like groups with chisel-like terminations); ramsdellite (seen as radiating black crystal aggregates to about 3 mm in length); todorokite (as silver-gray fibrous mats up to 4 mm in diameter); plate-like hematite crystal aggregates; as well as gypsum, fluorite, and celestite in a few specimens.

The primary mineral found in San Marcos geodes is quartz. The smoky variety is most common. Amethyst and colorless quartz occur less frequently and the amethyst often tends toward a smoky tint. Colorless quartz often appears dark due to transmission of the colors of underlying minerals. Secondary quartz was also observed as stacks of colorless quartz crystals on primary quartz or calcite. These stacks could be up to several millimeters high.

The minerals found in these geodes represent examples of those often found in Northern Mexican geodes, as well as some minerals (celestite and fluorite) not previously reported as geode microminerals. The source of the celestite and fluorite is possibly the underlying limestone since these minerals are often associated with that type of environment. Late stage hydrothermal activity could have transported the constituents of these minerals into the volcanic tuff above where recrystallization occurred within the geodes.

This site, in its embryonic stage, will be an interesting one to watch. The extent of the deposit is unknown and there is hope that it will provide collectors with an awesome suite of mineralogical wonders. As I stood on a nearby hill and looked down on the desert floor toward the mining area, I couldn't help but think this is exactly how the Las Choyas "coconut" geode deposit must have looked 45 years ago.

THE GEODES OF SALTO DO JACUÍ

Earlier I mentioned that in addition to the breath-taking Brazilian amethyst geodes, a second type of geode was found in Brazil's southernmost state. These are the geodes that are actually specimens of agate with a large hollow center. The agates range in size from fist size upwards to several feet in diameter, usually have a somewhat flat base, and are irregular in shape. Informally called agate "pots" by collectors, these agate geodes usually have a thick, banded agate rim (from one-half inch up to several inches or more) enclosing a hollow center of drusy quartz crystals. Minute rust red to black secondary iron and manganese minerals are often found delicately perched on the surfaces of the drusy quartz.

Although the possibility exists of being found anywhere within the agate-producing region of southern Brazil, most of these specimens originate in the Salto do Jacuí region. ("Salto" is Portugese for "jump" or "waterfall"; "Jacu" is a hen-like bird found in the local forest and "i" is an Indian word for "river.") The Salto do Jacuí region is about 31 miles south of Ametista do Sul's large amethyst geodes, or 190 miles northwest of Porto Alegre (a travel time by car of about four hours). The last ten miles of the road are not paved with asphalt. It's easy to find on a map of Rio Grande do Sul. Search for a big lake more or less in the middle

"Agate pot," Salto do Jacqui, Rio Grande do Sul, Brazil. (PHOTO BY DONNETTE WAGNER; BRAD L. CROSS COLLECTION)

of the map. The city lies just south of the lake. The lake was created for a hydro-electric power plant and mining activity of any type in this region is highly regulated by state environmental agencies due to sediment runoff. Because of large hillside mining excavations, as well as a number of other and larger contributing factors, the Jacuí River that feeds the lake runs a bright rust red color. For all you fast thinkers, yes, the gravel in the river is heavily concentrated with agate nodules.

The vast majority of the agate geodes from the Salto do Jacuí area are blue-gray in color. Specimens containing earthy shades of red, orange, or brown are likely found near the cutting centers of Soledade and Lajeado. Unlike many agate deposits I've visited, the agates do not tend to consistently blanket a specific area. Envision the occurrences much like large twisting snakes, winding in and out of the rolling hills, making it very difficult to construct adits and shafts. The geode-bearing basalt unit is sandwiched between an underlying 6- to 10-foot thick vesicular glass-rich dacite and a 16- to 23-foot thick vesicular rhyodacite. The volcanic complex often has frequent Botacatú Sandstone lenses. The agate geodes typically outcrop between 650 to 1,300 feet above sea level.

Exploration for an agate mine in this area is quite simple. Dense forests covering the gently flowing hillsides are traversed, the ground searched for small concentrations of agate float. These concentrations most likely indicate an outcrop of agate-bearing basalt. The host rock is rarely seen on the surface, but has instead been weathered to a bright rust-red soil. Initially, a 150-foot test line is dug to determine the general strike of the deposit. If the agate concentration is large and great enough, the hillside is excavated using modern earth-moving machinery. With each pass of the bulldozer's blade, men follow behind the equipment searching the ground for any agate nodule or geode. The bulldozed basalt is moderately weathered, so collecting with a rock hammer or pick is fairly easy. The specimens are then placed in a five-gallon bucket and later emptied into the back of a pickup truck. At the end of the workday, the truck is driven back to the farmhouse where the agates are stockpiled, some of the piles covering several acres.

Production amounts vary significantly. Any given mine may produce anywhere from several thousand pounds up to 500,000 pounds of agate per month. The specimens are then hefted, much like the Mexican coconut geodes, to determine their relative hollowness. Those that are determined to be geodes are not cracked open but are instead treated using lapidary techniques.

Most of these agate geodes have a somewhat flat base and a slightly domed top. The upper one-third of the geode is cut to reveal its inner sparkling beauty. Both pieces are then sanded and polished and ultimately sold as a pair. The bottom portion is somewhat reminiscent of a planting container, and thus the geode receives the informal name of an agate "pot." It's interesting to note that these agate-geodes are never found in association with the large amethyst geodes discussed earlier. They are separate and distinct mineral occurrences.

*Secondary minerals perched on a quartz druse make Ocos visually striking. Tres Pinaheiros, Rio Grande do Sul, Brazil. (*PHOTO BY DONNETTE WAGNER; BRAD L. CROSS COLLECTION*)*

Ocos

Within the dense rolling forests and intervening flat pampas grasslands of southern Brazil are found some of the greatest quartz treasures to be found anywhere. Among those many treasures are the Oco geodes. While many names have been given to different varieties of geodes found throughout the world, none of the names are perhaps as appropriate as those named "Ocos," which is Portugese for "hollow."

Ocos formed between 138 to 127 million years ago in amygdaloidal basalts that have now largely been weathered away, leaving accumulations of geodes in what is locally termed *cascalho*. The cascalho is a dark reddish-brown soil, derived from the extensive weathering of the Serra Geral Formation.

Ocos are found primarily in the steep hills of the Três Pinheiros (Three Pines) region. Located about 80 miles northeast of Soledade, the geodes are removed from the cascalho primarily through the use of hand tools – shovels and picks. In more productive areas, a backhoe is used to remove the hollow treasures from the shallow soil horizon. However, all of the work is not quite that

Tres Pinaheiros region where Ocos are found. (PHOTO BY VANDERLEI DOS SANTOS)

Ocos are mined directly from a dense basalt member of the Serra Geral Formation. (PHOTO BY VANDERLEI DOS SANTOS)

easy. Many of the Ocos must be mined from the tenacious basalt. With their relatively thin skin, extreme caution must be exercised in the mining process.

Not all Ocos are hollow. In fact, only 40 to 70 percent of the mined specimens will be worthy of a finished product. Once removed from the rich red soil, the geodes are placed in burlap sacks or buckets. When modern transportation isn't available, horses or mules are used to transport the geodes out of the area. Production volumes vary and are dependent upon the dollar and the demand for the specimens.

A pile of rough Ocos. Notice the somewhat flat shape of the geodes.
(PHOTO BY VANDERLEI DOS SANTOS)

Ocos range in size from one inch up to five or six inches in diameter. The average geode is comparable in size to a tennis ball and is many times prune shaped or somewhat flattened. When cut open at one of the cutting factories in the town of Soledade, clear drusy quartz will be found lining the interior against a backdrop of tan, brown, gray, or black. Their brilliant sparkle is reminiscent of tiny diamonds the size of granulated sugar. Only less than ten percent of the specimens will contain lightly colored amethyst. Many of the Ocos contain black, needle-like inclusions (probably goethite) among the minute, icy-clear quartz crystals. The particularly thin-skinned specimens are locally known as *casa fina* (thin shell) and will command a higher price. Most of the Ocos contain water and therefore are enhydros. However, few specimens are preserved as such and are instead cut and polished. Cut in half, polished, and offered in pairs, these geodes make striking decorative items and are affordably priced. It's not known exactly when Ocos were first marketed commercially.

Ocos are easily distinguished by their "rope-like" surface of chalcedony on both the inside and outside of the geode. When cut and polished, the rims display prominent white wavy or flame-like patterns similar to their Trancas cousins of Mexico and the desert roses of the southwestern U.S. Scientists have not been able to explain the formation of this common but unusual cloud-like chalcedony formation.

CHAPTER SIX

Geode Curiosities

— BRAD L. CROSS —

POLYHEDROIDS

*P*erhaps the most unusual geodes to ever been found are those known as polyhedroids – geometrically shaped geodes. These fascinating quartz geodes first made their appearance on the market in 1974. They were mined by Odwaldo Monteiro with his brothers in a poorly accessible part of Paraiba, Brazil. Paraiba is a small state on the northeast coast of Brazil. While it has many miles of coastline, the vast majority of the state is inland and consists of semi-arid farmland. The geode-producing area is gently undulating and covered by thick low scrub. In winter the area is swampy and in summer the temperatures can reach 112° F. Deep weathering has produced a thick deep red soil and it is in this soil that the polyhedroids are found. The host rock has been weathered away thus erasing most clues to their formation.

Externally the oddities appear at first sight to be single crystals ranging in size from a half-inch up to several feet across. They have cream, pale brown, to dark gray flat faces. The number of faces ranges from five to ten or more. The interfacial angles have no regularities whatsoever and range from 55° to 115°. There are no planes of symmetry.

Most all of the faces show two important features. First, they are not smooth nor are they striated, but frequently they are covered with thin silica sheets showing a regular pattern of small equilateral triangles very slightly impressed into the surface. The orientation of the triangular pattern varies in different parts of any one face. The pattern appears to represent casts of crystals which have

Polyhedroid geode. (Photo by Cindy Brunell; Gene Mueller Collection)

Fine polyhedroid geodes are now difficult to obtain. (Photo by Cindy Brunell; Gene Mueller Collection)

grown in a narrow, planar gap or joint. Secondly, many thin plates of very fine crystalline quartz, up to several millimeters thick, may be present adhering to the faces. In some instances, several of these thin scales may be present adhering to each other like a deck of cards. Many times, brown to reddish-brown limonite protrusions are found on the exterior and alternate with the thin plates of quartz. There is little doubt that the quartz plates represent vein-type infillings of planar fractures or zones. Most all of the edges are sharp and linear.

When cut in half and polished, several truly beautiful aspects are found. Most are hollow and are encrusted with colorless or white quartz crystals. Others are lined with blue, gray, black, and white-banded agate. Many times the banding alternates with crystalline quartz. Sometimes spectacular dendrites are formed between the layers of agate.

One theory of their formation is that they were formed in cavities between a latticework of calcite plates. Silica quite often replaces calcite and this type of replacement would explain their variety in number of faces, lack of symmetry, and range of angles. Essentially the geodes form filling the cavity just like the "gas bubble" geodes, except that the cavity has a geometric shape because of the crystal lattice.

The name, "polyhedroids," was given to the objects by J. L. Daniels because they are similar to, but not completely identical with, solids of many faces having no geometrical regularity.

ENHYDROS

Enhydros are geodes containing water. Frequently called "Nature's Water Bottles," the inner cavities are usually lined with drusy quartz crystals while the remainder of the geode is usually translucent chalcedony. The enhydros may be fairly round in shape such as those from Brazil or Mexico, or they may be shaped like fingers or bottles such as those from Tampa, Florida. Where the enhydros are known to occur, perhaps one in a hundred will, after careful scrutiny, prove to be enhydros.

Generally, the shell surrounding the water is of a translucent chalcedony that may not permit viewing the water. The water within may represent a residue from the original silica gel solution or the water may be recent and meteoric in origin, having been forced into the hollow cavity through pores or fractures by external pressure. Most enhydros contain less than an ounce of water with enough space left so that a moveable bubble is formed. Is it potable? Most water in the enhydros of volcanic origin is. But who wants to ruin one of these hypnotic beauties?

These water-filled agates make interesting specimens when ground and polished close enough to the water in the center that the movement of the water can clearly be seen as the enhydro is handled. Enhydros can very slowly,

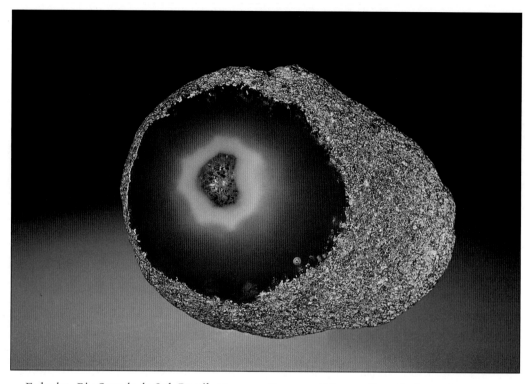

Enhydro, Rio Grande do Sul, Brazil. (PHOTO BY DONNETTE WAGNER; BRAD L. CROSS COLLECTION)

over the course of months or years, lose the water through cracks and pores of the enclosing agate. I've also seen it disappear because the freshly cut window was just too close to the pocket. In some instances, it is possible to restore it by long immersion in water. The fact that the water can be restored in some specimens may be taken as an indication that the water originally entered the cavity under external pressure. Though rarely beautiful, enhydros do make fascinating specimens.

Most of the enhydros currently found on today's market are from the Salto do Jacuí region of Rio Grande do Sul. These Brazilian specimens consist of blue-gray agate shells averaging two to six inches in diameter. When I went to the Otto Muller farm in Salto do Jacuí I was able to collect several nice fist-size enhydros within a half-hour's time. While I collected several dozen agates within this time frame, only two were proven enhydros. In one specimen, you could even hear the water sloshing around on the inside.

At this particular location, the agate-bearing unit was exposed on the side of a hill along the Jacuí River. Through the use of an excavator, the producing unit was terraced about six inches deep with each pass of the machinery. Hired help would follow behind the excavator, retrieving any agates that had been uncovered. The enhydrous geodes are then taken to a cutting factory where they are carefully cut and polished, providing a window to view the water within. There is no use of X-ray machines or any magical secrets in knowing where to cut the agate to present the best window possible. It's simply a matter of experience and luck. Each specimen is evaluated on a case-by-case basis before a saw cut is made. If a cut is made too deep and too close to the hollow pocket, the water is lost. Interestingly, the cut and polished specimens are submerged and stored in 55-gallon drums of water until their shipment to final destinations. This assures water in the geode is not lost prior to shipment.

In the early 1990s, a number of enhydros were discovered in Chihuahua, Mexico. These enhydros were 1.5 to 2 inches in diameter and were found in an altered ash flow tuff. The Mexican enhydros were not attractive in the least and were reminiscent of balls of caliche. The almost perfectly spherical shell consisted of a translucent chalcedony that was extremely thin and did not allow a window to be cut or ground. However, the better quality specimens contained enough space so that splashing could be heard when the rough specimen was shaken.

In the United States, besides the agatized coral from Florida, there are enhydros on beach locations in Oregon, Washington, and California. Handle them all carefully. Exposure to heat, freezing temperatures or rough handling can cause these natural water bottles to quickly lose their "magic."

Part Two

Gifts from Prehistoric Seas

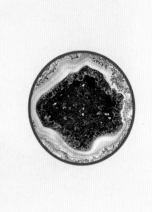

Agatized coral geodes are among some of nature's most beautiful creations.
(PHOTO BY JEFF SMITH; JEFF SMITH COLLECTION)

CHAPTER SEVEN

From Sediments to Glamour

— JUNE CULP ZEITNER —

No one ever stops to look at muddy sediments, but almost everyone stops to stare in wonder at a glistening diamond-like geode adorned with golden cubes and pink pyramids, a miniature Aladdin's Cave. Sedimentary geodes are among the most cherished treasures of mineral collectors everywhere and not for their beauty alone, but for the stories they tell of their birth and growth so long ago. Some people like to study the many and exciting variations of these geodes. Others like to specialize in a particular formation or locality. Others like to make microscopic studies of intriguing details that may be only dim to the naked eye.

One source states that the first use of the word *geode* with the modern connotation was in 1748 in a book by John Hill, *History of Fossils*. His book had a hand-colored plate with fine engravings of geode specimens. The word itself predated Hill, being used in 1619 as *earth-like* from the Greek word *Geo,* but not signifying a specific structure.

The word *geode* has been wrongly used in connection with vugs, nodules, thundereggs and concretions, all of which formed in other ways. The word *sedimentary* is a derivative of the old High German *sed* – sit or set, referring in this case, to materials settling in the bottom of a liquid – H_2O.

Sedimentary details are the results of the deposition of mineral sediments in prehistoric seas. The receding seas which covered a large part of middle North America left sandstone, mixtures of sand and shales, then sandy limestone, and then enormous deposits of limestone. The Mississippian limestone of around 250 million years ago is the home of fine and varied geodes in numerous well-known

*South Dakota's Englewood geodes have amethyst centers
often enhanced with pastel pink or peach. 4½ inches.*
(Photo by David Phelps; David Phelps Collection)

locations. Probably the best known of these plentiful occurrences is the Warsaw formation principally centered in a tri-state location centered by Keokuk, Iowa. In fact the geodes are often called "Keokuks." The formation was named by pioneer geologist, James Hall, for a small town in Illinois across the river from Keokuk. Technically the formation is a dolostone, since it contains more dolomite – $CaMg(CO_3)_2$ than calcite – $CaCO_3$. The Warsaw formation is fossiliferous in many places, so fossil hunters are attracted to these locations as well as geode hunters. In fact, one of the older theories was that sedimentary geodes had their start with silica layers filling cavities left by marine fossils. In some areas this has been found to be true.

SOME DEFINITIONS

There are various definitions of geodes. A current Webster's dictionary reads "A nodule of stone having a cavity lined with crystal matter." *Dictionary of Geological Terms* defines geodes as "Hollow globular bodies varying in size from an inch to a foot or more, and characteristic of certain *limestone* beds. Features are sub-spherical shape, hollow interior, a clay film between the geode wall and the enclosing limestone matrix, an outer chalcedony layer, and an interior drusy lining." That's a little better, but still misses the mark. Collectors who want to prospect for geodes following such a definition would miss some of the most intriguing, mysterious, and gorgeous geodes in the Americas! Even Leland Quick, writer, editor, lapidary and gem expert, wrote that the "most popular geodes are thundereggs." Not so! The much-admired thundereggs are only cousins! No definition leaves room for geodes shaped like bottles, fans, corals, brachiopods or starfish, or geodes containing crude oil, but these are all members of the sedimentary geode family.

Many sedimentary geodes started with an invertebrate fossil. 5 inches. Zeitner specimen.
(PHOTO BY DAVID PHELPS)

The blues in sedimentary geodes are lovely but are seldom as intense as the blues found in igneous geodes.
(PHOTO BY MILLIE HEYM; HEYM COLLECTION)

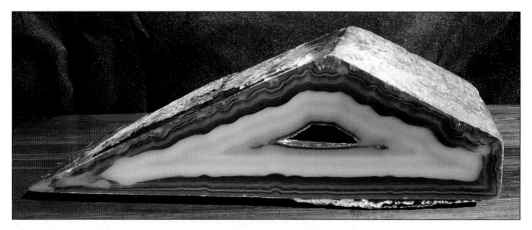

Colors in igneous geodes are often more vibrant than sedimentary geodes. Compare this polyhedroid with the Florida example above. Zeitner specimen.
(PHOTO BY DAVID PHELPS)

Many published geode definitions seem to be for geodes from igneous formations. They emphasize spherical and hollow and often agate. While the igneous geodes formed in gas cavities in magmas, sedimentary geodes formed in limestone cavities left by something usually rough and irregular. Most consist of a shell, hard and irregular, probably chalcedony, followed by cherty or agate bands surrounding a crystal filled or crystal lined cavity. Like geodes from igneous formations, the sedimentaries can have any number of inclusions. A large number of these geodes contain considerable calcium carbonate. Cavities in marine limestone may have been left by marine life that was literally "stuck in the mud." This is well illustrated by the geodes of Brown County, Indiana, shaped like a variety of sea life.

The oblate and semi-spherical geodes either have hollows, or sometimes in gradual growth *had* hollows and are now solidly full of quartz. Sedimentary geodes are composed of various successive layers during their formation. The "hollows" are the collectible ones.

Many geodes have centers packed with quartz crystals. 3½ inches.
(PHOTO BY DAVID PHELPS; JOE NONNAST COLLECTION)

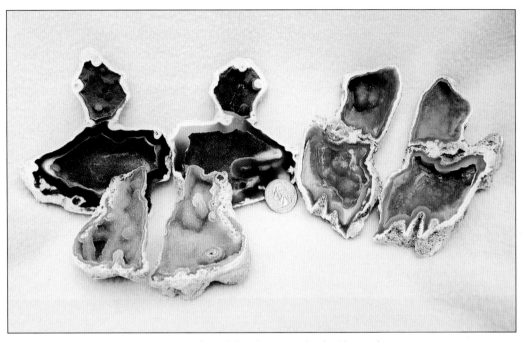

*Geodes are often defined as round. Florida geodes
and many others are not close to being round.*
(Photo by Millie Heym; Heym Collection)

The primary mineral in the formation of these geodes is quartz, either crystalline or cryptocrystalline (very minute crystals).

Geode walls may be thick or thin, smooth or rough. If a thick layered geode has easily distinguished successive layers, especially if there is a variation of color, the wall may appear agate-like and is polishable in case the geode is sawed rather than broken open.

Although sedimentary geodes are similar in many ways in most deposits, certain outcrops are often noted for specific inclusions, combinations of inclusions, crystal habits, colors, shapes, or percentage of solids. An example is the unusual number of pink calcite crystals found in northeast Missouri.

Geodes are widespread in Mississippian limestones or dolostones. Prolific examples occur in Iowa, Illinois, Missouri, Indiana, and Kentucky. Other localities are Tennessee, Alabama, Georgia, and Florida. These geodes have the following things in common.

1. They are discrete entities;

2. Exterior surfaces are harder and tougher than surrounding formation or environment; and,

3. Successive layers grow inward toward a central area which started as a cavity.

The outer skins are fibrous chalcedony, white, off white, gray, but sometimes stained by iron or manganese. The shell may be thin or may be thick enough to form the outer third of the geode. The concentric layers may vary in color and texture. In the majority of cases the greater part of each geode is quartz. Examinations have shown in the hollow geodes that the most recent crystallization is the nearest to the center.

Since dissolved minerals are present in enhydros (geodes containing water), it is possible such geodes could over time produce more crystal inclusions. Most rock contains cavities or vugs, but even a vug filled with crystals is not a geode, because a geode must have an encompassing hard outer layer with inward growth containing the crystals, and/or water, botryoidal lining or whatever else the shell encloses. A cave-like formation with calcite crystals covering the exposed walls is not a geode.

Although many Keokuk-type geodes are found in Mississippian formations there are exceptions. Two notable ones are the areas of Manhattan, Kansas and Wymore, Nebraska where similar geodes are found in the Permian, considerably later than the Mississippian.

THEORIES OF ORIGIN

Dr. David Hess summarized sedimentary geode origin theories in 1998:

1. Expansion of fossil shell and replacement of calcite by silica (Bassler, 1908);

2. Dissolution of interior of a concretion and replacement by silica and/or calcite, commonly leaving a cavity (Van Tuyl, 1916);

    a. Original calcite concretion theory with later replacement by silica and/or recystallization of calcite (Hayes, 1964; Sinotte, 1969);

    b. Original anhydrite concretion theory with later replacement by silica and/or calcite (Chown, 1974);

3. Silica gel theory – solution fills cavity (Robertson and Brooks, 1994; Pettijohn, 1957).

Hess also expresses his view that more than one theory might be correct depending on the location, the formation, and the classification.

I have always felt that since I have several excellent recipes for my chocolate cake, Mother Nature would probably have a lavish supply of recipes for her masterpieces.

VARIATIONS

Geodes from various localities become easily recognizable with study, because the minerals and circumstances of growth differ, granting a distinctive quality to those found in a specific locality. I found iron-stained geodes in Kentucky and Tennessee but not in Iowa. Most Brown County, Indiana geodes are identified by their shapes. The Missouri geodes have more calcite, especially pink calcite, than those of Illinois. South Dakota geodes have rough rounded bumps on the exterior with a thin gray skin which appears baked on. The many variations are due to the mineralization of the location of the earliest pre-geode concretions, the limestone or dolomite host, and the former placement of geodes in the bottom and upper layer of the formations.

While people usually think of a geode as a "hollow rock," Hayes, Sinotte, and Finkleman are among the scientists who agree a cavity need not be present. Only that there was one. Although many sedimentary geodes are called spherical, most are elliptical and many are found with a slightly flattened pole. In addition, some geodes are fossil shaped and some are quartz paramorphs after shells or corals. (The latter occur principally in Florida.)

The size variations of sedimentary geodes are extreme. Some measure only an inch or so in diameter and have shells thin enough to float, while others may be several feet across and weigh 200 pounds or more. The variations of colors and inclusions are as great as the variations in shape, size, and weight.

Exterior colors can match any color of the spectrum. Interior forms and shapes also run the gamut. There are stalactites and stalagmites like little caves. There are bubbly mammilary and grape-like growths. Cubes, pyramids, wafers, needles, prisms, octahedrons, scepters, hexagons, and more add to the splendors of the interiors.

Chalcedony and quartz are the most common components of geodes. South Dakota geode. Black Hills Museum of Natural History. 4 inches. (PHOTO BY LAYNE KENNEDY)

Dr. David Hess listed these known geode inclusions as well as the very rare and unconfirmed inclusions in geodes in an unpublished paper he wrote with Mike Sandstrom in 1998:

Quartz, chalcedony, calcite, dolomite, ferro-dolomite or ankerite, goethite, aragonite, smithsonite, malachite, pyrite, marcasite, chalcopyrite, sphalerite, hematite, kaolinite, pyrolusite, barite, gypsum (selenite), jarosite, millerite, violarite – polydymite, celestite, pyrrhotite, smythite, petroleum, water, and gas.

A new discovery in Argentina yields this sedimentary geode with an interior of botryoidal opal.
(PHOTO BY DONNETTE WAGNER)

Quartz includes amethyst, rock crystal, rose quartz, milky quartz, smoky quartz and quartzes tinted green, red, or blue.

Mark Sherwood adds rare szomolnokite plus unconfirmed jamborite, honessite, retgerite, zaratite, wurtzite, todorokite, romanchite, tenorite, magnetite, aurichalcite, sulfur, galena, copiapite, and fluorite.

In the past the following have also been listed as inclusions: siderite, byssolite, hornblende, chlorite, azurite, rutile, dickite, and opal. These are named in papers by one or more of the following: Van Tuyl, Hayes, Sinotte, Fleener, Tripp, Schaub, Hurlbut and Tarr.

CHAPTER EIGHT

Wonderful Warsaw

— JUNE CULP ZEITNER —

Sedimentary geodes present a huge and intriguing variation in shape, size, color, weight and inclusions. They are found in mountains, hills, valleys, prairies, plains and along rivers, lakes and seas. Most geodes which defy discretionary descriptions are sedimentary.

The best known locality in the Americas for sedimentary geodes is commonly called *Keokuk*. Actually Keokuk is the name of a southeast Iowa city on the Mississippi River where the geode bearing Warsaw Formation is exposed for some sixty miles in all directions. The so called "Keokuks" are also found in Missouri and Illinois with similar ones being found in Indiana and Kentucky.

The Keokuk area is the center of the Mississippi River Arch that reaches from St. Louis to near Madison, Wisconsin. In this region Mississippian formations can be seen on either side of the Mississippi River. The Mississippi River Arch is sandwiched between the Wisconsin Arch and the Ozark Uplift. The average elevation of the Keokuk area is 600 feet. Two glaciers invaded this area, the Kansan and the Illinoian. Glacial drift still covers parts of these states.

PHYSICAL CHARACTERISTICS AND INCLUSIONS

Keokuk geodes, like most sedimentaries, are mostly quartz. They are more or less oblate, irregular and rough on the exterior. Some exteriors have multiple large bumps, similar to an oval balloon with hives.

95

This twin-chambered Keokuk geode has a common chalcedony, quartz, and calcite combination. 7-inch Zeitner specimen. (Photo by David Phelps)

Goethite is included in this Iowa geode.
(Photo by David Phelps; David Phelps Collection)

The hollow interiors are a source of delight for collectors. The mystery surprises may be colors, textures, lusters, or inclusions of any of about thirty minerals. Hopes of finding new or more beautiful inclusions are what keep field collectors busy.

The possible inclusions in these natural packages are calcite, pyrite, marcasite, chalcopyrite, barite, dolomite, goethite, pyrolusite, selenite, hematite, aragonite, smithsonite, malachite, siderite, kaolinite, byssolite, hornblende, chlorite, millerite, ankerite, sphalerite, jarosite, fluorite, petroleum, gas, and water. All these plus quartz varieties, chalcedony, rock crystal, amethyst, and citrine.

The sizes of the tri-state "Keokuk" geodes are as small as a prune and as big as a winning county fair pumpkin. The majority are between 3.5 inches to 6.5 inches. The big ones reach 30 inches in diameter. Sizes tend to be similar in separate occurrences. The geode deposits also host concretions and fossils, although most have only geodes. Most of the geodes occur in the Lower Warsaw and a few in the upper Keokuk Formation which is just below the Lower Warsaw.

The limestones of the Warsaw Formation were formed during the Mississippian period around 350 million years ago when Iowa, Illinois, and Missouri had steamy tropical climates. The Mississippian is topped by the Pennsylvanian, when the coal beds were formed. Often when hunting for geodes a thin coal seam is a sign to look a few yards lower for geodes.

Brown and pink calcite gives this Missouri geode a sharp color contrast. 1¾ inches.
(PHOTO BY DAVID PHELPS; DAVID PHELPS COLLECTION)

Fragile acicular crystals of millerite in an Illinois geode are a rare occurrence.
Dr. Hess purchased this at a flea market. (Photo by Larry Dean, WIU)

Origins and Locations

How did the geodes form? "That's a good question" is an often heard glib answer. Of course the real answer is no one knows for certain. There are several theories, some of which seem preposterous and several of which seem probable or possible. Personally I like to think Mother Nature had more than one recipe for geodes in her cook book.

One theory claims they were formed after anhydrite concretions on the sea floor. J.B. Hayes wrote that they are epigenetic, that is they were formed after the enclosing sediment hardened. A popular theory has been that geodes were formed in cavities left by marine life such as sponges, crinoids, blastoids, horn coral, brachiopods or sea urchins. In fact, Dana's 4th edition of *Manual of Mineralogy* published in 1895 mentioned the sponge cavity theory.

R. S. Bassler, writing a Smithsonian paper, went along with the fossil theory using corals and brachiopods as his examples of the origin of formation cavities.

As the limestone of the Keokuk-Warsaw-Salem became dolomitized in the sea bottom, the dolostone, high in calcite, was replaced by silica making the geode shells pseudomorphs. This theory was accepted by many mineralogists, however each made certain exceptions in regard to the predominant geodes of their own specific sites.

For Keokuk-type geodes, many scientists now agree that the silicification of the anhydrite nodules occurred contemporaneously with the precipitation of the anhydrite nodules at or near the sediment water interface in response to calcium saturation. The high proportion of calcium occurred because of the dolomitization and sulfate reduction. The source of silica may have been sponge spicules. When the sulfate rich cores of the nodules were gone voids were left for the geodes.

Scientists now think that the geodes are quartz pseudomorphs after anhydrite, a calcium sulfate, usually associated with gypsum. The anhydrite nodules were said to have formed on tidal flats in extremely hot conditions

Yellow ram's horn dolomite is the "gold" in this Iowa geode. 4½ inches. (PHOTO BY E. EARL SMITH; E. EARL SMITH COLLECTION)

but metamorphosis started in warm deep seas. There have been plenty of absurd explanations of the origin of geodes also, including petrified potatoes, peaches and pumpkins. An acquaintance showed me one he insisted was a petrified bird's nest, he even showed me the botryoidal formation which he called "eggs." Sinotte pointed out that a few early geode finders associated them with meteorites! (Impossible!)

Other more reasonable theories will be found in succeeding chapters. The final verdict is not in.

In the Keokuk area the Warsaw Formation can be seen in many spots along the banks and bluffs of the following rivers and streams: In Iowa, the Des Moines and Skunk Rivers, Bear Creek, Mud Creek and Weaver Creek; In Illinois, La Moine River, Railroad Creek, Tyson Creek, Larry Creek, Crystal Glen Creek, Dallas Creek; In Missouri, Fox River and Weaver Creek. Some of the cities and towns in geodeland are: Iowa – Keokuk, Burlington, Salem, New London, Lowell, Mt. Pleasant; Illinois – Hamilton, Carthage, Colchester, Dallas City, Quincy, Pontoosuc, Warsaw, Niota; Missouri – Canton, Alexandria, St. Francisville, Kahoka, Wyaconda, Winchester.

I once bought a geode at a show in Cedar Rapids, Iowa with a label giving Delta, Iowa as the location. If this is so, it is well outside previous areas. A friend, Ruth Dean of Des Moines, told me she found geodes along the Skunk River near Des Moines. I have not confirmed either place, but I do think new geode occurrences will continue to be discovered and among them will be wonderful deposits like the recent finds on Highway U.S. 61 in Missouri.

This is a scenic area with lots of big old deciduous trees – maple, oak, ash, hackberry, together with stately pines and lots of flowering shrubs and neat storybook farms. The spacious two story country homes with their lilac hedges and picket fences lend a nostalgic charm. The many small towns were built over a century ago. Although appearing sturdy and durable, most are losing population, while a few that grew to 10,000 or more are attracting new industries and experiencing new growth.

Geodes are the official state stone of Iowa. South of New London is Geode State Park, one of a handful of such parks in the United States that honor a native rock, mineral or fossil. There are abundant geode beds in the park. (No collecting allowed.)

The picturesque site is along the Skunk River. The Indian name "Chiquaqua" sounds better. Comprised of over 1,600 hilly acres, the park has forested acres, a lake, and well kept facilities for visitors. The park land was purchased by nearby

Saddle-shaped yellow dolomite frequently occurs in Warsaw geodes. 5 inches.
(PHOTO BY E. EARL SMITH; E. EARL SMITH COLLECTION)

communities wanting to preserve a part of geological history and heritage of the state. State money created the lake for fishing and swimming and built roads and erected signs. Scientist Ben Hur Wilson noted that the depression era W.P.A. constructed the picnic shelters and restrooms using native rock.

Important in promoting and publicizing the park were the Smiths of New London, six miles from the park. The Smiths named their lapidary equipment company "Geode Industries." Their early tumblers were big favorites of lapidaries and their introduction of Vibra Sonic tumblers transformed the industry. They also wrote several guide books for tumblers, and went all out to help geode bound Iowa visitors. They stocked and advertised choice specimens.

Observant travelers will notice many signs in the area that they are in geode land. In the town of Mount Pleasant near the state park is an attractive footbridge made of handsome geode specimens. Some of my Iowa friends told me they have found geodes farther west of the park along the Skunk River. A geode location is near Lowell, and along tributaries to the Skunk River. Inquire locally.

The "gold" gleaming in this geode is pyrite, the "fool's" hope, but nevertheless a precious geode. Collected by David F. Hess at Bear Creek, Iowa, in 1995. (PHOTO BY LARRY DEAN, WIU)

Brown calcite with dogtooth calcite fill the cavity of this St. Francisville, Missouri, geode. David F. Hess bought this one at the St. Francisville Rock Shop.

The type locality for the Warsaw geodes is not at Keokuk but across the Mississippi in Illinois at the town of Warsaw. The lower geodiferous bed of the Warsaw Formation is 40 feet thick. The wide upper layer is hard blue-gray clay. The type locality specimens often have dramatic inclusions such as pink dogtooth calcite with cubes of pyrite, or golden saddles of dolomite with bronzy sphalerite.

There is no other way to describe the density of the geode population in the Keokuk area than for me to describe my first experience in geode wonderland. It was in the early spring some 40 years ago. We had spent the winter digging geodes in Florida and were on our way back to our South Dakota headquarters. The weather was just what one hopes a perfect Spring day will be, so we decided to make a stop at the St. Francisville geode location, which had been described to us by a friend from the Des Moines Lapidary Society. On one side of the stream were rolling green hills covered with the largest purple violets I had ever seen. Across the stream was a rather steep blue-gray cliff, literally studded with geodes looking exactly like they had been laid in place by a master stone mason, like a

The combination of quartz, kaolinite, calcite and dolomite is found in all three of the original Warsaw formation states. (PHOTO BY E. EARL SMITH; E. EARL SMITH COLLECTION)

stone wall designed by a professional. Beneath each row was a duplicate row, but slightly to the right or left so the quite uniform geodes did not touch. There were 12 rows of tempting balls of white quartz.

Albert paused to take a few pictures, while I happily began to collect the geodes which had already fallen into the high water of northern Missouri streams in early spring. Albert started working with hammer and chisel at the bottom where geodes were easily released while I carefully hefted each one to determine if it might be hollow.

Sadly, it wasn't long before we had the limit we had imposed on ourselves because we already had a load of rocks in our International Travelall and R.V. So after a violet scented picnic we started homeward, with me secretly hoping we didn't spot another likely looking field trip area, because I was worried that *one* more rock would give us a flat tire, which was a frequent event in those days. (We had the flat anyway!)

TYPES OF SEDIMENTARY GEODES

Geode specialist Steve Sinotte, divides the geodes into lower and upper classes and then subdivides them according to type. He calls the lower Keokuk type S, type C and SC. The upper are KS or KC.

Type S has a chalcedony shell with a recrystallized layer of quartz between the shell and the interior. They have a ratio of quartz to calcite of more than 50 percent. The similar SC geodes have a calcite/quartz ratio of more than 50 percent. S or SC geodes are the only ones with kaolinite in the cavity.

Type C geodes have chalcedony shells, but no other quartz. Re-crystallized calcite replaces the quartz interior lining. The brownish calcite is sometimes iridescent and is fluorescent. Type C geodes are rare.

Type SC geodes have brown or pink calcite but also some re-crystallized quartz. Kaolinite may also be present. Doubly terminated quartz crystals are prized additions.

Type KF are the geodes which Sinotte says *resemble* horn corals, blastoids or crinoids. KF geodes are rare in the Keokuk area, with the only finds limited to Railroad Creek, Illinois. The geodes are usually lined with excellent quartz crystals and have no calcite.

Type KC which Sinotte calls "vughs" (vugs) would not be included as geodes according to current definitions.

ILLINOIS GEODES

There are considerable differences in the appearances, sizes, and inclusions of geodes in different localities. Hamilton, Illinois "Keokuks" are a case in point. Gray's Quarry in Hamilton has horn coral shaped geodes. Railroad Creek nearby has geodes shaped like echinoids. Snowball geodes are common in Illinois but not in Missouri. Millerite occurs in Illinois but not in Iowa. Niota, Illinois is the site for petroleum geodes. Malachite is found more often near Dallas City, Illinois than elsewhere.

Illinois has plentiful pyrolusite inclusions near Hamilton. Dewdrop diamond geodes and blue chalcedony lined geodes from near Niota are other favorites for collectors in Illinois.

Calcite and dolomite on quartz add focus to this irregular cavity of a Hamilton, Illinois, geode.
(PHOTO BY LARRY DEAN, WIU; DAVID F. HESS COLLECTION)

Barite rosettes, chalcopyrite and sphalerite are rather attractive inclusions from such Illinois locations as Perry, Pontoosuc, and Crystal Glen Creek.

The type locality for Warsaw formation geodes is Geode Glen where a large exposure on Warsaw Creek revealed numerous pink calcite enhanced geodes.

A 4-inch geode from Hamilton, Illinois, has silvery white quartz and yellow dolomite. Zeitner specimen.
(PHOTO BY DAVID PHELPS)

MISSOURI GEODES

I have heard of Keokuk-type geodes found as far from the type locality as Hannibal, Missouri. I have not been able to verify this, but I like to think Mark Twain was familiar with them.

The geodes of Missouri are thought to be the most beautiful overall. Several locations have for many years produced a large number of aesthetic specimens. Numerous great finds have been made at Alexandria and St. Francisville. Other locations are near Wayland and Kahoka. Some of the largest ones have been found in this area and there seem to be more hollows.

Barite crystals are frequent residents of St. Francisville geodes. The tabular crystals may be clear, yellow or blue. The blue may be caused by clay or kaolin. My best St. Francisville special has blue barite against what appears to be "citrine" – goethite stained yellow quartz.

Stunning recent finds in the tri-state area are the iridescent brown geodes of Fox Hills, Missouri. Purchased from Betty Scheffler by David F. Hess.
(PHOTO BY LARRY DEAN, WIU)

Transparent yellow barite adds glamour to geodes from several localities.
(PHOTO BY LARRY DEAN, WIU; DAVID HESS COLLECTION)

From Alexandria come some spectacular pyrite lined showpieces – glistening golden cubes. Iridescent marcasite over botryoidal chalcedony occurs near Kahoka.

Also, among the Missouri premiums are pink calcite rhombs, snowy kaolinite, and dark striated pyramids of sphalerite, sometimes sprinkled with pyrite. The sphalerite with pink dogtooth calcite lining makes a stunning specimen.

The vivid red-shelled geodes of the La Grange road cut on U.S. 61 are highly unusual. Kenneth Vaisvil writes that the rare geodes of the Canton road cut near La Grange are lined with pink dogtooth calcite crystals with uniform characteristics. Vaisvil's mention of intense blue chalcedony shells near Canton makes me want to call my travel agent at once. The recent discoveries in Missouri are truly amazing.

QUARTZ IN GEODES

Quartz is the main attraction of most Keokuks. The chalcedony shells are all quite similar, only varying in size, shape and texture. The most common linings are drusy quartz crystals facing toward the hollow. Crystal inclusions may be short prisms with pyramidal hexagonal terminations. Etched, striated, twinned or doubly terminated crystals occur. The vitreous luster and the translucency are assets. Smooth grape-like botryoidal chalcedony often with splendid inclusions, is another quartz variation.

Another kind of quartz crystal found in geodes is called phantom. The euhedral crystals exhibit growth lines or "ghosts" of successive stages of formation.

Most of the quartz is colorless or white, but some of it is colored by inclusions or mineral coatings or stains. These added colors can be yellow or reds from goethite and hematite, black from sulfides, blue gray from clay, or rarely lavender from ferric iron.

This quartz crystal "snowball" geode was found in the Betty Sheffler Geode Mine in Missouri. (PHOTO BY KEVIN FOSS; KEVIN FOSS COLLECTION)

Quartz is a tectosilicate, 7 in hardness and 2.65 in density belonging to the hexagonal system of crystals. Although acidic it is chemically passive. The chalcedony which forms the outer shell of most geodes is fibrous, thus adding to the strength of the geode.

One of the Keokuks most desired by collectors is called "snowball," an appropriate name, since these natural spheres of white chalcedony seem to be glittering with frost crystals. The spheres sometimes almost fill the geode cavity. Looking like Christmas decorations, the balls are attached to a small area of the lining. Some are the size of a ping-pong ball, others like a golf ball and once in a while even a baseball. The crystals on the spheres face outward almost touching the inward pointing originals.

Another very desirable collectors' geode is aptly named "dewdrop diamond." The cloud-white chalcedony lining is liberally decorated with shining clear doubly terminated dipyramidal crystals. In variations the quartz "diamonds" may be slightly smoky. Opening one of these rarities is indeed like opening a treasure chest. Many of this kind of geode have been found across the river from Keokuk along Glen Creek, near Hamilton, Illinois. Some have also been found in Iowa and Missouri.

CALCITE IN GEODES

Calcite is the second most important mineral of the sedimentary geodes. Calcite may be pink, white, brown, yellow or black. Calcite crystals belong to the scalenohedral division of the hexagonal system. Crystals are lustrous and have numerous habits often making identification confusing. Some of the shapes are called nail head, barrel, and dogtooth. Fluorescent calcite crystals turn yellow, orange, or red under black light. Thin calcite coatings over quartz may fluoresce a brilliant white. Phosphorescence is rare, but not unknown. Many geodes do not participate in fluorescence phenomenon.

Some geodes contain iridescent calcite, vibrant purple, crimson and turquoise. Other calcite inclusions may have a silky luster or a pearly luster. Calcite can display colorful incandescent flashes when geodes with clear calcite crystals are turned in the sunlight. Geodes can also hold doubly terminated calcite crystals as well as phantom crystals. Clusters of peach colored calcite crystals can be shaped like flowers, fans, or wings.

Pink calcite crystals are showy against the white quartz of this Missouri geode. 5 inches.
(Photo by Larry Dean, WIU; David Hess Collection)

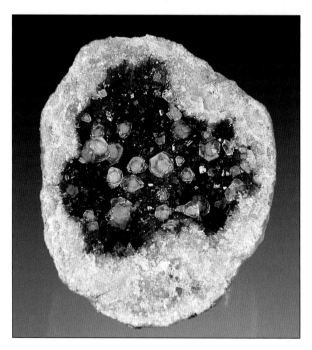

Calcite provides a bright contrast to the dark iridescent quartz. 3½ inches.
(Photo by Jeff Smith; Jeff Smith Collection)

Engineer Kenneth Vaisvil in *Rocks and Minerals* described a recent discovery in Lewis County, Missouri. The locations are on the highway construction which will eventually take Betty Scheffler's geode mine. Highway work has unveiled an exceptional number of splendid iridescent calcite lined geodes. Blue, purple, crimson, gold and green iridescence enhance the rich brown calcite. Reticulated marcasite and red chalcedony outer shells are novel variations.

BEAUTIFUL COMMON INCLUSIONS

Kaolinite is a component of numerous geodes. An alteration product of feldspars, in the geodes it is often pearly white. It is also a component of other geode minerals causing the pastel pinks of calcites and the baby blues of barite. Geode outer shells incorporate kaolinite as a crystalline powder. Kaolinite was a very early inclusion in geodes, while selenite was the last.

Pyrite lined geodes are spectacular when miniature golden octahedrons line the cavity or when brilliant cubes enhance pink calcite or blue chalcedony. Besides the druses and clusters, pyrite also appears as hair-like capillary crystals in transparent host crystals, and rarely as in radiating stars.

Fancy geodes with shimmering iridescent siderite coatings in the hollows are almost opalescent. The thin film varies in splendor according to the color of the mineral it coats with its flashes of blue, green, and violet.

Stellate groups of slender aragonite crystals are scarce in Keokuks but the star-like inclusions are among the most showy. Under short wave ultraviolet light aragonite fluoresces yellow or red. It may occur with an equally fluorescent calcite.

Rhombohedral dolomite crystals decorate geodes with interesting shapes and a wide range of color, mostly from ferrous iron. Dolomite colors may be pink, peach, yellow, taupe, pale lime, ivory and gray. Aggregates of curved crystal faces stand out among the straight lines of other crystals. Little saddles have a pearly luster.

Dolomite is found with quartz or chalcedony as well as calcite and barite. Its twin carbonate, ankerite, occurs in geodes also but less frequently.

Goethite has several assignments in geode decor. The prismatic crystals appear as blades, druses, tufts or clusters. Colors are mostly browns, yellows and blacks. It also occurs as coatings on botryoidal quartz. Some dark goethite has a metallic luster. Lustrous black needles in water-clear calcite or quartz emulate millerite.

An ankerite psuedomorph after dolomite adds a bright accent to this Illinois geode. 5½ inches. (PHOTO BY E. EARL SMITH; E. EARL SMITH COLLECTION)

Thin transparent blades of selenite are common in Keokuks. These fragile late bloomers, glassy and clear, add a light and delicate touch to any geode. The sharp slender crystals are often perfect examples of this variety of gypsum. Small selenites are prevalent in sedimentary geodes, but fine crystals up to three inches long are rare surprises in a few Keokuks.

The above inclusions are some of the most interesting and often the most collectable of geodes from the tri-state Warsaw formation. The excitement of finding and opening geodes in the hopes of finding something new, more puzzling, more beautiful, or extraordinary lures one on. But failing a record find, the one who opens each mystery box is thrilled to be the first in the world to see it.

There are also geodes with several cavities instead of just one. These cavities are usually all different in size and shape, but well separated by their chalcedony walls. I have counted six individual cavities in one "multiple" geode. (Often called a double.) These multiple cavity specimens are often heart shaped or like a puffed up figure eight. The crystal linings may be different in each pocket. The chalcedony shells fit together like a jigsaw puzzle.

Mineralogist and author Ben Shaub mentions that sedimentary geodes sometimes carry some of the limestone matrix, when broken out of the formation. Once in a while a collector interested in the complete story of geodes will collect a geode surrounded by matrix. The matrix dolostone of an outcrop near Kahoka, Missouri was so overcrowded with geodes that many touched each other, altering their normal growth, so that they were not even close to being spherical, and for some the shapes were just plain weird.

Shaub also pointed out that the carbonate structure cone-in-cone may be adjacent to geode surfaces. He wrote that the late forming crystal inclusions were superb for micromounters.

At least one of America's most influential mineralogists, Willard Roberts of South Dakota, primary author of the

Jeff Smith found this rod of goethite covered with goethite needles like a bottlebrush in appearance, an exotic geode incusion.
(PHOTO BY JEFF SMITH)

most indispensable mineral book *Encyclopedia of Minerals*, was inspired at age 5 by the wonder of geodes. His father had given him a glittering pyrite specimen for Christmas so Bill started looking for more pretty "rocks." He noticed all of the geodes in his Iowa grandfather's rock garden and around the yard and flowerbeds, so he began appropriating his choices with the smiling approval of his grandparents and parents.

The geodes of the Warsaw Formation are found in yards in the towns and countryside throughout the tri-state area. Residents have found many ways of using and displaying their durable stone heritage. Handsome stone walls are studded with geodes. Geodes are used in foundations, flower beds, wishing wells and planters. Native robins, wrens and cardinals splash around in baths made of halves of large geodes. Some geodes, whole or in halves, become paperweights or door stops. Fountains in city squares or parks are inlaid with geodes. Best of all, numerous long time residents of geodeland have collections of these famous geodes, nicely prepared, and well labeled and displayed.

Choice examples of Keokuks are seen at club and federation mineral and gem shows in geode-rich states. Mineral dealers in the area are able to fulfill the wishes of visiting collectors or those who order by mail. Books show glamorous pictures of splendid examples. (The cover of one of my own books features a geode portrait.)

In the days before color photography we had to be satisfied with the verbal descriptions of inclusions as interpreted by various writers. For example Iowa collector and writer Ken Borschell describes one of his favorites this way, "...about the size of a baseball and thin shelled. Quartz crystals within the cavity are exceptionally tiny, almost microscopic." He goes on to describe the colors and luster, but no word description is adequate for those unfamiliar with top quality Keokuks.

Kate Steinbrenner of Des Moines had an esthetic geode lined with blue botryoidal chalcedony accented with pink calcite rosettes. Verne Waterman showed me a heart shaped 7-inch diameter geode with tabular translucent yellow barite crystals against almost cubic, highly iridescent brown calcite. Steve Sinotte showed a geode lined with hundreds of gleaming crystals of pyrite, so bright that some called it the "gold geode," not bothering with the usual prefix "fool's."

Some of the shapes of inclusions that make certain geodes the most outstanding and collectable are fans, scepters, pyramids, roses and rosettes, pagodas, spires, domes, icicles and spruce trees.

Keokuk geodes are found in natural science museums around the world. In the mid-1800s, A.H. Worthen, Illinois geologist and geode enthusiast, sent barrels of fine geodes to museums in Europe, Asia, Australia, Canada and, of course, the most prestigious museums in America.

Perched in this Keokuk, Iowa, geode is a showy crystal of sphalerite.
David F. Hess purchased it in the Tudor Rock Shop in Illinois.
(PHOTO BY LARRY DEAN, WIU)

In 1970 there were over 200 geode sites recorded in the tri-state Keokuk region. Considerably more have been added since then and new sites are discovered on a regular basis because of erosion, excavations, and construction projects. This does not add to the number of places to collect however, since many of the old areas are worked out.

Some record setting geodes got their "15 minutes of fame" and more. One of the early Warsaw giants was pictured in the *Mineralogist* in 1944. Two feet in diameter, the crystal lined geode was found near the Fox River on the Otho Hagerman farm just out of Wayland, Missouri. The owner was a fossil collector. Russel Kent found a 300 pound Wayland geode. Dr. Wells Sinotte of Keokuk had

Keokuk geode. (PHOTO BY DONNETTE WAGNER)

a 27.5-inch diameter monster weighing 500 pounds from near Fox City, Missouri. It appeared to be full of frosted grapes. Verne Waterman told about seeing a man near Kahoka dig out one that measured 28 inches across. Judging from the number of pumpkin sized geodes I saw in tri-state farm yards, there are probably many more over two feet, and maybe even over three feet. Strangely most of the larger ones have been found in Missouri. (The Show Me State!)

Each locality and each subdivision has its own characteristics. The melon sized giants and most of the pink calcites come from Missouri. The oil filled ones are at home in Illinois. The strangest shaped ones are in northern Iowa near Mason City. Of course there are always exceptions, and that only adds to the excitement.

KEOKUK'S NORTHERN COUSINS

An interesting and surprising geode location that has never had the attention showered on the Keokuk area is in northern Iowa, and also many miles farther west in the state than the previous locations. All these geodes, unlike the Keokuks, have few inclusions. While in the Keokuk area a minute proportion

116

have been found with any marine life resemblance, the geodes of the northern area not only have astounding marine life forms of crinoids, corals, and brachiopods, but they have characteristics indicating their true fossil origin. In these strange geodes, botryoidal linings occur more frequently than in southern deposits. Another difference is in size. The southeastern geodes are far more obese than the northern relatives. A nine-inch diameter is considered very big in the north, unlike the twenty plus inch diameters of the south.

The northern area geodes have been found along the Iowa River and its tributaries near Steamboat Rock and Iowa Falls. Other locations are near Chapin and Sheffield. The north most outcrops are not far from Mason City and near a State Park at Clear Lake and along the Winnebago River.

Some of the collectors in these areas have found specimens that made them suspect that marine *plants* may have left the voids which were filled with chalcedony and quartz. Most specimens replicate marine invertebrates.

Banded agates in neutral colors found near Manchester and Delhi on the Maquaketa River also show fascinating marine life origins. These northern occurrences have not been explored thoroughly and in-depth studies have not been finished.

Except for state parks the area is all private land. Permission must be obtained even for collecting at abandoned quarries. Sometimes geodes may be found along old roadways, railroads, or exposed cliffs near bridges across rivers and streams. Recent excavations also hold promise.

Ken Borschel wrote of finding fossil-like geodes he identified as "silicified stems." He adds that they are reminiscent of stemmed geodes he found near Imlay, South Dakota. (I had not heard of these, but that doesn't mean they don't exist.) Could his "stems" be geodized sections of crinoid stems such as those of Indiana? More about this later.

Borschel reported that the geodes from the Chapin and Sheffield sites were composed of a hard greenish greasy-feeling material. This was not observed by other writers. He theorized the color could be from decaying vegetation. (Is it possible the greasy feeling could come from nearby oil?) He also found brachiopods extremely well preserved. The upper part of the Keokuk formation directly below the lower geodiferous Warsaw is also known for geodes. It is the upper Keokuk formation in the north that has the fossil-type geodes.

The most common in these exposures are those shaped like crinoids, blastoids, and horn corals. In the Keokuk region the only source of this kind of geode is at Hamilton, Illinois.

Fossilized geodes is not an accurate name for these unique specimens because the geodes themselves are not fossilized. The terminology changes in other areas, but the varied names only show how little attention has been paid to these divergents. Sedimentary geode collectors should reexamine all their specimens with open minds to see if any differ from what we all think of as typical Keokuks. If some of them prove to be puzzles to the collector, he should show them to other collectors and scientists to help the studies of these neglected objects of the Mississippian period.

It is possible that new locations will be found in the future in northern Iowa or other states that will reveal more about these strange fossil-type geodes. Perhaps even now a graduate student is planning his Doctor's thesis on this very subject.

How to Find and Enjoy Geodes

A popular fee basis geode site is Betty Sheffler's mine. Her shop is in front of the mine near Alexandria, Missouri. It is in a unique building constructed of brick with a wonderful geode facing. There are 60 tons of geodes and other fine specimens encouraging some rockhounds to loiter outside before entering the well-stocked and well-organized rock shop.

Contact an official in Alexandria to find out if the mine is still open. Thousands of people have found their own prized geodes at Betty's and became ardent collectors after viewing her geode gallery.

Although people from all over the United States and the world have been among Betty's customers she says as far as she knows the most famous and perhaps notorious geode collector was Ferdinand Marcos, former president of the Philippines. The state has planned to build a new highway through this property. Thousands of fine geodes have been found by hundreds of happy people.

Members of the Keokuk area warn that early Spring is not a good time to collect geodes. Rivers and streams are at their highest levels when all the snows have melted and run off. Late May and early June are good times and the autumn seasons are usually excellent. Residents of the region say they are often able to collect in mid-winter or mid-summer if the weather happens to be just right.

It is always best to have a hat or eye shade and since good collecting is often near a river or stream, waterproof shoes or boots are advisable. Tools include Estwing rock pick, chipping hammer, gads and chisels.

Usual rock bags are not suitable since two or three good geodes could fill the bag. A big plastic pail or burlap bag or a sturdy box are better choices. (Line the box with plastic.) Safety goggles are needed to chip the geodes out of the limestone matrix. Unless a collector is going to a familiar spot, adequate maps are best friends.

Since the local gem and mineral club often take field trips to productive sites, it is a good idea to write or call these clubs to find out when such trips occur, and if non-members can attend. If the club's insurance bans non-members, often an individual member of the club volunteers to help visitors. Better yet join a club. After all these clubs are proud of their geodes and like to share their treasures, to trade specimens, and to be invited to rock and mineral locations outside of their immediate region. Clubs are connected regionally and nationally. Clubs are listed in the annual May Buyers Directory of the *Lapidary Journal,* a Primedia publication, or visit the AFMS website at www.amfed.org for information about the Regional Federations and clubs.

Geodes can be collected at a city park in Warsaw, Illinois and another park in Hamilton, Illinois. There have been fee basis collecting sites at Wayland and St. Francisville from time to time. Inquire about possible new ones from Missouri clubs.

OPENING, PREPARING, AND SHOWING SEDIMENTARY GEODES

When opening geodes judged to be hollow, it is best to prepare the exterior first. The finder always hopes the interior will be "a sight to behold" so it is too late to clean the outer shell if wishes come true, and rare and fragile inclusions are found prior to exterior preparation.

First remove soil, plant roots, or anything that might have clung to the geode. The outer shell is hard and tough so knives, gravers, styli, picks or dental tools can be used. Next brush the geode briskly with a stiff bristle or wire brush. This is followed by a bath, but before the bath be sure there is no hole or fracture that would allow water to seep inside.

Put the blemish-free geode in warm water with a favorite detergent. Let it soak for a while. If the shell is stained by rust or vegetation, Clorox may be added to the water. Scrub the geode with a brush or a real sponge. Follow the bath with a thorough rinse.

Many of the specimens from Niota or Nauvoo, Illinois, may be stained with tar. Soak such specimens in kerosene or gasoline, but do this outside in a safe area, to avoid toxic fumes. Diesel oil or carbon tetrachloride are good cleaners too. Use outdoors with care. After removing stains in a monitored fresh air environment, scrub geodes with hot water and detergent. A final dip in denatured alcohol may improve the outer appearance, but it may not be possible to remove all of the ancient stains on some of these geodes. Some collectors prefer to leave the stains.

Iowa geode expert Steve Sinotte recommended a solution of ammonia and distilled water for a final cleaning for most geodes. This gives a slight glow to quartz which is more smooth or which appears bubbly. I sometimes spray a geode having a handsome exterior with Pledge® and buff it off at once. A few geodes are slightly iridescent on the exterior. Give them a gentle blow dry – nothing else.

It is best to open the Keokuks without using the diamond saw. Cutting oil may ruin some inclusions and a clean smooth cut ruins the natural appearance. One way to open a geode is to use a hammer and chisel after a careful examination. Many Keokuks have a slightly flattened bottom, and my husband thought fragile inclusions might be attached to that bottom. He would decide which part of the geode might make the most aesthetic miniature cave, but he never purposely broke a geode exactly in half. He usually guessed correctly as to the beauty of the larger portion he wanted for the feature, the "Aladdin's cave." If the geode was longer than it was high he would stand it on end for a more interesting appearance. Using an aluminum "pencil" or a pale washable ink marker he would draw a line around the circumference. Then with his hammer and a newly sharpened chisel he would score the line, going around the geode several times tapping lightly. In some spots the chisel would go through the shell deeper so more work was done at these spots, and even a slightly firmer tap or two would finally open nature's surprise package.

An alternate way and probably faster is the use of a pipe cutter as described in Chapter 3. For larger geodes which tend to have thicker shells and fewer delicate inclusions, I have seen experienced hobbyists drive awls or gads through the shell in a few well chosen spots. But without expensive tools, the only way for a novice collector to learn is to read, ask questions, and then try.

If a geode is to be used for a specific purpose, for example to hold a candle, a saw can be used to cut off a shallow part. Self sticking felt can be put on the bottom, and the candle can be waxed in place inside in the center.

There are wire and plastic stands on the market that can be adapted for geode display. (Often they are made for antique saucers or plates or to show fine porcelain in a china cabinet.) It is not difficult to bend 14 or 16 gauge wire into geode holders. The wire only needs to support the bottom and part of the back.

The stand should be simple so that it does not detract from the specimen. For small geodes, save the round cardboard circles from Scotch tape. Spray them flat black, and they will make useful and inconspicuous stands for geodes about three or four inches in diameter.

For larger geodes, take some heavy aluminum foil and crush it around the bottom and lower back of the geode until the geode is in the best position and is held in place and supported by the foil. Spray the holder with matte finish paint either black or off-white.

A good geode display should be labeled with such information as location and inclusions, and maybe name of the collector. The label should be rigid enough that it will not be warped by heat, and small enough that it will not detract from the geode. (Usually only the serious gem and mineral collectors bother to read labels.)

A thick shelled geode with a botryoidal lining or a lining of only drusy quartz, for example, can be polished. First grind the edges until they are smooth and even, unless the geode has been sawed with a good diamond saw. Stuff the cavity with crumpled damp newspapers and put the halves on a vibrolap. Sand first with coarse grit, about 100, then 220, 400 and 600 followed by buffing with tin oxide.

For children, putting tiny plastic or ceramic animals in the hollow will attract and hold their interest. If a geode looks like a miniature ice cave, adding a wee toy polar bear would give the teacher a chance to give two lessons at once.

If children have a chance to see how a geode is opened, a good game is for them to guess simple things about what the interior might hold, such as color, crystal size, number of shapes, whether there will be a smooth bubbly look, or crystals, and if there will be surprise. (Like additional colors and shapes that resemble something familiar.)

Children also enjoy making little stands for geodes with cardboard, putty, clay, or wire. They can grow small flowering plants in geode halves for Valentines Day or Mothers Day.

Keokuk-type geodes are found in many countries. A friend found some in Panama. I saw a display of minerals from Peru that contained sparkling quartz/chalcedony geodes similar to those from Missouri.

Geodes from Georgia's Bartow County found in barite pits, are lined with high quality quartz crystals.

Other states with Keokuk-type geodes are Kansas, Nebraska, Oklahoma, Texas, Indiana, Ohio, Kentucky, Tennessee, Alabama, and Alaska.

My Thoughts About Midwest Geodes

I have seen the price of aesthetic geodes rise rapidly over the years. About twenty years ago I bought a "dewdrop diamond" geode from Missouri for $20.00. Now a similar one might be $100.00 or more. Although geodes can be ordered by mail or the Internet, it is best to see before you buy. Good moderately priced specimens are usually available at Midwest shows or swaps, or at auctions or silent auctions sponsored by clubs or individual collectors. Barter is still a good way to acquire choice specimens. Buyers who look for rare inclusions or exceptional beauty in rock shops in the Keokuk area may pay more, but they also may find specimens not available elsewhere.

Undoubtedly new locations for sedimentary geodes of the Keokuk variety will be found in the future. Probably new inclusions or new combinations, or colors or shapes will be discovered, studied, and identified. Possibilities are exciting as thousands of geodes emerge from the captivity of their matrix.

Hopefully there will be new fee areas for collecting as farmers and other landowners realize that geode localities are a source of income. I encourage clubs in geode areas to get together to buy or lease a geode-producing site so they will have a legal place for collecting and for other clubs.

Collectors who find something rare, unique, or unidentifiable in a geode should get help solving their problems, and then get the new information out to scientists and fellow collectors.

Several states have notable sedimentary geode deposits in addition to the tri-state area of this chapter. In the Midwest are Indiana, Nebraska, Kansas, and South Dakota. In the Midsouth are Kentucky and Tennessee, and in the South are Georgia, Florida, and Texas. However, I think most people with an interest in minerals when they see a fine geode will exclaim, "Look at the Keokuk!"

CHAPTER NINE

Strange Geodes of Indiana

— JUNE CULP ZEITNER —

Many years ago I saw an amazing feature exhibit at a Midwest show so unique that there was always a fascinated crowd around it trying to get a better look. I stayed in that crowd until I got close several times to stare in awe at something I had never seen before, and had not even read about. What the large case apparently held were quartz fossils, marine inverte-brate fossils. There was a fat gray snail, an obese five-pointed fossil resembling a starfish, a ballooning crinoid stem, a perfectly distinct brachiopod, a blastoid, and more. They looked like fossils but they all looked similar. Something was not right. No wonder! The inconspicuous label said "Brown County Indiana Geodes." Geodes?! I failed to locate the owner and failed to find anyone who could tell me more about these geodes, but I soon learned the Brown County location – Beanblossom Creek. So in the third edition of my book *Midwest Gem Trails* (1964) I added the Brown County Beanblossom Creek location for geodized fossils, mentioning that some paleontologists thought the geodes only had a "resem-blance" to fossils. The geode connection was a controversy in the 1960s and in a way is still a mystery.

As my interest grew through the years, I determined to find out as much as possible about the Indiana geodized fossils. I found very little in print so I contacted the Midwest Federation State Director for Indiana, asking her if she knew anyone in any of the Indiana clubs who had a collection of these strange geodes. To my sur-prise, and probably to hers, she was the one most interested in these geodized fos-sils and the one acknowledged to have the largest and best collection. Margaret Kahrs sent pictures to me of her collection and later some specimens to study.

*The three geodized fossils from Indiana, found by Margaret Kahrs, are a 2½-inch brachiopod, a 1½-inch gastropod, and a 2¾-inch crinoid. (*PHOTO BY DAVID PHELPS; ZEITNER COLLECTION*)*

Margaret lives not far from Brown County where exposures in the Ramp Creek Formation along Beanblossom Creek have the geodes. Other Ramp Creek exposures with geodes are in nearby Lawrence and Monroe Counties.

The geodes vary in size from about the size of a big walnut to the size of a pumpkin. On average their largest measurement is from three to four inches. Some of the quartz outer shells are quite smooth except for distinct features of the original marine fossil. Nevertheless many retain the shape of the fossil but not the characteristic details. These specimens have rough surfaces and you can only see other distinguishing marks here and there. In fact, were it not for the superior smooth, easily identifiable fossils, many or most of the lesser specimens may never have been recognized as geodized fossils.

Mineral inclusions in the geodized fossils are not as common as in other types of geodes, but they include about the same species as other geodes, for instance those of the Warsaw Formation. The most common inclusions are calcite, gypsum, dolomite, goethite and pyrite. Others are sphalerite, marcasite, ankerite, celestite, hematite, siderite, anhydrite, strontianite, aragonite, millerite, galena and kaolinite. Many geodized fossils have been found in the Ramp Creek Formation in Indiana, but the greatest number of them have not been opened; and those that have been cut or otherwise opened may not have been examined for inclusions by scientists. The collecting of these fossils has been chiefly by amateurs. Interested amateurs may be mistaken in reporting inclusions, for example, one admitted to me that what he reported as colorless calcite turned out to be selenite. A mineralogist called needles of goethite "millerite." There are a few occurrences of bitumen or asphalt in the Ramp Creek geodes like the oil geodes of Niota, Illinois.

An exceptional size for a geodized gastropod is this 4½-inch specimen from the Jeff Smith Collection (PHOTO BY JEFF SMITH)

The mystery to me is that there has been so little attention paid to the geodized fossils by professionals. It seems as if there should be lists of known localities, how the geodes originate, genera represented, what genera are most abundant, how to recognize them, whether all are Mississippian, and many other questions.

One geologist told me the largest geodized fossil is about the size of a softball, but my late friend William Allaway, an advanced amateur fossil collector, wrote in 1960 that the largest were like pumpkins. He wrote that he had examined geodized crinoids, brachiopods and corals. Margaret Kahrs's specimens show

From Barite Hill, Indiana, come attractive geodes with crystals of pale yellow barite on quartz and calcite.
(Photo by Jeff Smith; Jeff Smith Collection)

his observations to be true, however scientific papers prove to be illusive.

Just what are some of the known fossils that have become geodized? Apparently only a small percentage of them have features for positive identification as to species. Among those I have seen are brachiopod *Spirifer crawfordsvillens*, a pudgy 3-inch specimen, smooth surfaced with lots of fine parallel lines and *Athyslamm eilosa*, which still looks exactly like a brachiopod in spite of all it has gone through. Brachiopod *Spirifer Keokuk* is named for an Iowa cousin. A detailed horn coral is named *Triplophyllum dalei*. A cephalopod is *Muenstroceras parallelum*. Most of those treasured by their finders are merely labeled crinoid, blastoid, gastropod, pelecypod, coral, bryzoan, or rarely echinoid or starfish.

Various parts of crinoids are found in most of these geode localities. They are quickly spotted by fossil collectors who are familiar with crinoids. The calyxes and columns make distinctive geodes. The five-point symmetry of some bases and stems are hard to miss. Flowerlike heads are often only partial and are rare. Some blastoids have been mistaken for parts of crinoids.

Some geodized shells have a part of another shell or a small barnacle clinging to them. Since echinoids may already resemble geodes it is hard to classify them. (Barnacles are crus-taceans not mollusks).

One of the less common inclusions in Keokuk type geodes is millerite. A noted location is near Bedford, Monroe County. The sprays, tufts, mats, and webs of millerite are often spectacular, shimmering like fine gold wire. The millerite occurs with calcite, dolomite and barite. R.M. Ley in *The Mineralogical Record* compares this occurrence to Keokuk, Iowa, and Hall's Gap, Tennessee. The location is no longer very productive.

I have an ammonite that is obviously a chalcedony geode. Golden in color and smooth in texture, it is somewhat translucent and shows a perfect outline with a living chamber and graceful central swirl. I have two gastropods that could not be mistaken for anything else, one a nice quartz snail

Horn corals are uncommon.
This large example is 5½ inches.
(PHOTO BY JEFF SMITH; JEFF SMITH COLLECTION)

with swirls to the center of five elegant continuous wharls. The other one looks like a cone shell. It is 2½ inches long and ¾-inch in diameter. There are six diminishing layers to the tip. Both are dexterous – right-handed. It would be satisfying to have the genus and species of each, but I consider these good specimens because they are not posing as fossilized invertebrates. They are posing as what they are, geodes with a fossil heritage.

A delicate spray of millerite takes center stage in this Keokuk-type Indiana geode.
(PHOTO COURTESY JEFF SMITH; COLLECTED BY B.B. EVANS)

The chances are that many geodes from the Mississippian locations in Indiana, Iowa, Kentucky, and Tennessee, which now look quite ordinary, could trace their family tree back to Mississippian invertebrates. I feel that if we look closer we may find clues to prove this.

The Ramp Creek Formation used to be a member of the Harrodsburg Limestone Formation but now is classed as a separate formation just below the Harrodsburg. It is at the top of the Osagean Series of the Mississippian and correlates with the Warsaw of Iowa.

Geodes are found along and in creeks, in road cuts, construction sites, and in and around dams and lakes. Beanblossom Creek flows through Monroe County, where Lake Lemon is an important feature. The lake has been a site of geode finds. Beanblossom Creek empties into the White River. Lawrence County is immediately below Monroe County, bordering Brown County. These three have been primary sites.

Dr. Gary Lane observed that these geodes were recognized as being of fossil origin in the 19th century. He also points out that the geodes are similar to those of northern Iowa in the Mason City area. He describes the prevailing theory of the origin of the geodized fossils this way, " Originally deposited as gypsum in a hot saline environment and then replaced by quartz and calcite." Dr. Norm R. King says the fossils served as seed for the growth of the geodes but that it is not clear how they formed. He feels that almost any of the almost complete Mississippian marine fossils or complete parts of such fossils could be the start of geode growth. (A complete part could be a crinoid base or head.) He quotes another theory about the origin and growth.

"It seems it helps to have a hollow space surrounded by shell or other skeletal material. There is an apparent osmotic flow of ground water into this space. This water brings in dissolved mineral matter (largely silica) that precipitates here when it reaches a high concentration and also replaces the original skeletal material, which is calcium carbonate. The volume increases during this process and the outer margin expands. The distortion and preservation quality during the

Margaret Kahrs found the 2½-inch yellow chalcedony ammonite and 2-inch gastropod.
(PHOTO BY DAVID PHELPS; ZEITNER SPECIMENS)

An ardent geode collector checks out
a popular roadside site in Indiana.
(Photo by Jeff Smith)

'explosion' would preclude precise identification of the species involved."

He further explains that this explosion distorts the details of the fossil as the calcium is replaced by silica, making positive identification difficult in most of the geodized fossils.

Still another theory dates from the early 20th century. Scientist J.B. Hayes said geodes started with diagenetic concretions beneath the water/sediment interface at the floor of the Mississippian Sea. They were formed by the precipitation of dissolved calcite over sites of high pH resulting in the decomposition of marine life that had burrowed into the sediment and died. The warm shallow seas were alkaline and dolomitization and silicification occurred followed by cavity formation and induration. R.S. Bassler noted that geodes in Kentucky and Indiana plus a very few in the Keokuk region originated in the cavities of brachiopods, corals, and crinoids.

Dr. William Sanborn thinks the geodes are replacements of concretions formed in deep water where biological material, such as shells, sank to the bottom and attracted calcite to form a biolith around the material. The concretions became incorporated in rock layers by sedimentation. Acid laden water may have dissolved the calcite, which was replaced by silica.

The unreal "designer" geodized fossil is a crinoid calyx from Heltonville, Indiana. 1½ inches.
(PHOTO BY JEFF SMITH; JEFF SMITH COLLECTION)

Steve Sinotte made no attempt to explain the origin of geodized fossils although he pictures a few. He discusses several theories of geode formation and shows four good examples of geodized fossils in his book, one from Indiana and three from Hamilton, Illinois. The blastoids and the gastropods are excellent choices, but from his 77 black and white and colored plates, I think that as many as five more are also geodized fossils.

Iowa writers Muriel Menzel and Marilyn Pratt accept the theory of the formation of some geodes found in sedimentary formations as starting with a limy concretion precipitated around a nucleus of a

marine animal. The growth of the concretion stopped when it was coated with silica. Then the inner calcite of the concretion dissolved leaving a cavity within a quartz shell. The writers collected at Hamilton, Illinois where most or all of the geodized fossils of the Keokuk Warsaw region have been

The crinoid stem is 1¾ inches,
a rare specimen.
(PHOTO BY JEFF SMITH;
JEFF SMITH COLLECTION)

found. But such geodes are the exception there – at least those that are apparently of fossil origin.

Paleontologist Dale Douglas believes most sedimentary geodes of the Midwest originated from fossils mainly brachiopods, corals, crinoids, snails, and clams. Dr. R.D. Rarick, Indiana Geological Survey, leaned toward that theory and also observed that since the evidence of fossil nucleus soon disappeared in the long complicated process of geodization, it would be difficult to prove that *all* these geodes started with a fossil.

Geodized fossils, once called "puffed fossils" are found in many stages of transformation. Those that are found showing their fossil generally are the exception.

Once in a while a distinct geodized fossil is found in an Indiana Pleistocene formation after being pushed out of place by glaciers.

Sometimes concretions are mistaken for geodes and there are some similarities, particularly when comparison involves larger concretions. Concretions are different from geodes in the manner of growth. Geodes are hollow or have been hollow and the growth from the outer shell is inward. Concretions grow outward from a central nucleus. The nucleus may be flora or fauna. Such concretions are solid not hollow. They lack the symmetry of layered growth.

Dale Douglas called geodes from the sedimentary formations "fossil pseudomorphs." The process of geodization proceeds at a snail's pace. Although it may have started in the mineral laden sea, the seas of the Mississippian gradually became the Mississippian Period in the middle of a continent like the Warsaw of Iowa and the Ramp Creek of Indiana. When the everchanging geodes are exposed at the surface of a rock formation the process stops. Douglas thinks that in some places the geodization process is still continuing.

Douglas has a series of geodized crinoid stems that look like pieces of sugar cane or corn stalks. He arranges them to show at least seven stages of the development of the crinoid geodes. The first are smooth all around and the same diameter throughout. From the third stage on, the stems start swelling and exhibiting stretching and wrinkling. Most came from the Middle Mississippian near Bloomington but two came from the glacial drift near Middlebury. Hollow silicified crinoid stems are found in Montgomery County.

Indiana petrologist S. Greenberg says the geodes of Indiana are the source of the greatest variety of minerals in the state. The Ramp Creek and the Edwardsville Formations of the Osage series of the Mississippian have shown the most species. Geodes are common in many counties in Indiana, but the geodized fossils have become known in only a few. Besides Brown, Lawrence and Monroe, geodes occur in Washington, Jackson, Morgan, Warren, Owen, and Floyd Counties.

An area of Jackson County is noted for a different geode phenomena – geodes, usually with rather thin shells and loose crystals inside, variously called rattle rocks, or rattlers. A well-known area for these is in Jackson County just south of the Muscatatuck River near Highway 135. Many of these geodes have beautiful quartz crystals and interesting inclusions of numerous minerals.

Ohio is not as geode rich as Indiana; nevertheless it does have a few locations for geodized fossils similar to those from its neighboring state. Some of the geodes resemble sponges. Others are like corals, brachiopods, pelecypods and blastoids. The sponges were not mentioned in the Indiana references. The most

A geode crinoid stem from Heltonville, Indiana, illustrates how the original stem has been blown up. 2 inches.

(PHOTO BY JEFF SMITH; JEFF SMITH COLLECTION)

popular Ohio location for collectors has been near Hillsboro in Highland county. Brush Creek in Jackson County and May Hill in northern Adams County are other geode locations.

Geodes patterned by ancient marine life are exciting and intriguing, but they raise plenty of questions, some of which can be answered and some that remain mysteries. For example, why only the Mississippian formations? And why only marine invertebrates? Why not marine plants? Why not Eurypterids or sea scorpions, or trilobites, or agnaths – "jawless fish"? Why not algae or perhaps sponges? Why are some of these fossils so beautifully preserved in their metamorphosis into geodes while others are barely distinguishable? The sad truth is we don't know. Maybe there are some ancient plants and animals involved in the geode genesis that we just have not discovered yet. These geodes are relatively new to the geological sciences of mineralogy and paleontology. It was only a little more than 60 years ago that many scientists were accusing amateur discoverers of having overdeveloped imaginations.

RECOGNIZING GEODIZED FOSSILS

Not long ago I saw a display of fifty Keokuk-type geodes at a club show. The only label read "QUARTZ GEODES". I examined each as carefully as I could through the glass. I looked at shape, texture, pattern, line, uniformity, non-conformities, and considered my first impression of each, and my reasons for doubts. Out of the fifty I felt sure that there were six that were geodized fossils, and noted two or three "probables." I asked a casual passerby if she thought any of the geodes reminded her of sea shells, and one was only on my list of probables. If only the first six I had decided on were true geodized fossils, I wondered how many other geodes in this case developed in distorted cavities or maybe developed from fossils in a turbulent environment.

Each geodized fossil area has its own characteristics. Some localities have an abundance of a specific fossil class, order, or family. Or certain localities have few large specimens while nearby large ones are the rule. The Ramp Creek formation covered hundreds of thousands of years, so the environments of all these fossils could not have been equal. As we discover more and learn more about geodized fossils, we probably are in store for many surprises.

Not all the geodized fossils are hollow. Like the Keokuk geodes, many are full of quartz. Those that are very distinct in showing their origin, like one of my brachiopods, are hollow and lined with sparkling inward facing quartz crystals the same as other hollow sedimentary geodes which do not have a fossil seed. Lighter weight, thin shelled geodes no matter what the shape, are more likely to be hollow.

The geodized fossils that are difficult to recognize look different enough from common geodes that many finders make stabs at naming their genesis and come up with labels like petrified corn cobs or walnuts, pine cones, or sugar cane.

Indiana also has Keokuk-type geodes. This beauty features ankerite.
(PHOTO BY JEFF SMITH; JEFF SMITH COLLECTION)

Or they mistake a geodized coral for petrified honeycomb. Another factor leading to misidentification is that there are pseudofossils, rocks that have a cursory resemblance to some once living plant or animal. Water, ice, wind and sand are good at shaping lookalikes. The most common pseudo fossil is "petrified potatoes."

Since the geodes of Iowa, Missouri, South Dakota and Kentucky are Mississippian, it seems like Mississippian formations in other states might be good places to explore for more geode deposits. Places that might be hiding interesting geodes are mountain exposures of the Mississippian. A friend, mineralogist Tom Loomis, found a geode in the Black Hills, not a quarter of a mile from my home. None had ever been reported there. Another friend, Vince Henderson, gave me one from the same locality looking like an albino Englewood amethyst geode. Another possible exposure could be during excavations for building a dam. As co-leaders of Operation Rockhound, my husband and I had our caravan stop at a new dam near San Luis Potosi, Mexico, just to see what the large exposure revealed. What we found were geodes all over the place.

This Indiana birdbath reminds me of the one my father-in-law made with orange South Dakota geodes.
(PHOTO BY JEFF SMITH)

One reason the geodized fossils and other geodes have not received much attention from professionals is because geologists and mineralogists are trained to look for mineral deposits large enough to be of significant economic value. Geodes are appreciated mainly by amateurs and museums.

Displays of geodized fossils are seldom seen, and many amateurs do not even know of their existence, but these geodes are well known in southern Indiana. They are common in rock gardens, in borders around landscape features. They are embedded in concrete walls. They outline lawns. They are piled up in farmer's field corners with rocks of glacial drift. Heavy ones wind up as doorstops. Children play with them, maybe even imagining themselves at the seashore as they pick up corals and gastropods. And some people like Margaret Kahrs collect them, study them, and label them.

CHAPTER TEN

A Few Bonuses

— JUNE CULP ZEITNER —

Perhaps the greatest rival of the Keokuk area for geode locations and sheer numbers of specimens is Kentucky. The Bluegrass State has rugged mountains and thick forests in the southeast portion. Cumberland Gap in the Appalachians was once considered the gateway to the "West." The awe-inspiring scenery along the many rivers reveals inviting geode country. The Pottsville Escarpment divides the Cumberland Plateau from the Bluegrass Region and the Mississippian Plateau. The Bluegrass area itself is divided between the Inner and Outer Bluegrass. The shaly area bordering the Plateau and the Bluegrass for many years has been known as "The Knobs." The Pennyroyal Plateau is carved of Mississippian limestone. In the limestone country some of the streams are diverted underground. The major geode-rich counties form a loose semi-circle around the Bluegrass, reaching southward into Tennessee. Notable geode deposits have been discovered in Estill, Madison, Lincoln, Rockcastle, Jefferson, Adair, and Lyon Counties.

There are several well-known geode-producing areas in Kentucky. A Lincoln County site is Halls Gap near Stanford on U.S. 150. Another is not far from Louisville in Jefferson County. I recall names of several Knobs in this large area called "The Knobs," but all I could find on a current map is Gap-In-Knob south of Louisville. The most visited site in this geode intensive area is Muldraugh, once referred to as Muldraugh's Hill. (Incidentally these geode treasures are found near Fort Knox, known for another mineral treasure.) In the same district is Buttermold Ridge, noted for fossils as well as geodes.

Keokuk-type chalcedony/quartz geodes are abundant in Kentucky. 6 inches.
(PHOTO BY DAVID PHELPS)

The Warsaw-Salem Formation is prominently exposed in Adair County. In most of these counties geodes were found when the state was first populated prior to 1792, but pioneers considered the geodes a nuisance and seldom examined them or opened them. They were in the way in farmer's fields, road projects, excavations and quarries. They were sometimes used in foundations, walls, bridges, or for doorstops. Some Abraham Lincoln scholars have listed a Kentucky geode in Lincoln's small rock collection.

The Mississippian formations of Kentucky and Tennessee have abundant sites for sparkling and varied geodes. Some of the geode-bearing formations are thousands of feet thick. Valleys and streambeds down slope from a familiar formation, the Warsaw of Iowa, Missouri, and Illinois, are literally choked with geodes. Here the formation is called Warsaw Salem. Many locations have the common irregular quartz-chalcedony "spheres" of the tri-state Warsaw region. Knobby and gray on the exterior, the geodes have many inclusions, mainly quartz, calcite, selenite, dolomite, pyrite, hematite, and goethite. More rarely sulfides such as chalcopyrite, sphalerite and galena are found. Fluorite is present in small amounts. Occasional inclusions are millerite, rutile, and manganite.

In some areas the geodes have distinctive rinds, stained yellow, orange, or rusty red by iron. Sizes in specific deposits also tend to be similar, for example an area known as "the pumpkin patch" tells it all. Fine agates are found with the geodes of Cumberland Plateau.

Most Kentucky land is private. There is one National Forest in the eastern part of the State named for Kentucky pioneer Daniel Boone. There are also many abandoned quarries, but permission must be asked for these, as well as private farms or acreages. Great places to hunt are around lakes or reservoirs, along rivers or streams. And along road cuts particularly country lanes. Look for blue-gray Mississippian limestone outcrops. An area not far from Danville near Herrington Lake in The Knobs, has a pretty creek loaded with geodes and lined with geodes at times. A stream near Lexington is also a geode hunter's paradise.

This iron stained Kentucky geode was found on soil rich in kaolinite near the Green River.
(PHOTO BY MARINA ESTES)

Tom Estes gets ready to wade across the Green River to see if hunting is better on the other side.
(PHOTO BY MARINA ESTES)

SILICIFIED ANHYDRITE'S FOSSIL FAMILY

H.L. Barwood and N.R. Shaffer of the Indiana Geological Survey mention Ramp Creek equivalents near Louisville. It is in these deposits of silicified anhydrite nodules that geodized fossils are found. Also geodes from such deposits may contain unusual minerals. Ramp Creek and similar limestones are dolomitic. Fossils are abundant in such Mississippian limestones. Frequently fossils are found which are only partially turned into geodes. Some are geodized fossils and others, which may have started with fossils, may reveal their origin only by extensive testing.

Paleontologist Ray Bassler of the Smithsonian reported that in some strata almost every fossil cephalopod can also be rated as a geode. He also mentioned some cephalopods are multiplex geodes since each separate chamber may have its own inward facing crystal lining. In some instances fossil shells embedded in limestone are replaced by silica only in exposed parts. The parts of shells covered with limestone are still calcium! Some of the Kentucky geodized fossils are Ordovician instead of Mississippian, but those of "The Knobs" many times are washed down from higher strata. Some geodized fossils have even been found in the city limits of Louisville.

Warren H. Anderson of the University of Kentucky noted that some of the fossil crinoids and blastoids have enlarged to softball-sized geodes, and that brachiopods and gastropods were common nucleii for other geodes, some of which became pyritized. Becker and Day thought the majority of Knobstone geodes could be traced directly or indirectly to crinoidal origin, followed in numerical order by brachiopods. Complete replacement of the calcite by silica may leave a

A puzzling geode from Poosy Ridge, Rockcastle County, Kentucky, is partially coated with kaolinite. 7 inches. (PHOTO BY DAVID PHELPS; ESTES COLLECTION)

141

perfect representation of the fossil, on the other hand it may destroy the details of origin.

Specimens show that brachiopods are more likely to have life details well preserved.

Small crinoid parts are more apt to be misidentified or to go unnoticed entirely. The Keokuk Warsaw Salem Formation of Barren County, Kentucky is a prime location for crinoid-type geodized fossils.

Geodes of Estill County, Kentucky are often difficult to identify, as they may have few reminders of their fossil origin, and fewer traces of anhydrite, one of the founding minerals. A popular attraction in Estill County is Natural Bridge State Resort Park, in a scenic part of east central Kentucky.

Near Boone and Mount Vernon in Madison County in the Bluegrass Region are more geode sites. Geodes have been found in road cuts near the college town

This coral head geode is from the Green River in Kentucky, a well-known site for similar geodes. 6 inches. (PHOTO BY DAVID PHELPS; ESTES COLLECTION)

142

The abundant coral head geodes of Kentucky come in various sizes, shapes, and colors. (PHOTO BY MARINA ESTES)

of Berea. Some of the most beautiful Kentucky geodes I have seen came from Lincoln County. Pink calcite, blue celestite, golden pyrite and lavender fluorite have decorated these.

Like the Indiana geodes, the Mississippian geodes of Kentucky are thought to have originated from anhydrite nodules precipitated at sediment water interfaces in the ancient seas in response to calcium saturation. The abundant calcium came from a combination of dolomitization and sulfate reduction of magnesium sulfate brines, according to geologists Barwood and Shaffer. The silica source may have been sponges. The removal of sulfate cores left the voids characteristic of geodes. (The anhydrite origin was found by Hayes in 1964.)

Many of the Kentucky geodes belong to the Sanders group. The quartz of the geodes varies from fibrous and microcrystalline to the showy inward facing euhedral crystals of the interior. The earliest silica mineralization in these geodes is spherulitic, probably replacements of anhydrite as well as possible crystallization from a gel state. Twisted fiber chalcedony is characteristic of geodes.

As in the Iowa, Missouri, and Illinois Warsaw geode areas, the geodes of Kentucky are widely sought by field collectors, rock and mineral shops and their customers, and schools and museums. The publicity engendered by significant finds results in at least a modest addition to the state's economy. Certainly the fascinating geography of Kentucky and the great names like The Knobs, Gap-In-Knob, Halls Gap and Bluegrass should capture the interest of school children.

TENNESSEE

The Mississippian geode locations of the Tri-State, plus Indiana, and Kentucky, extend into Tennessee, Alabama, and Georgia. Formations equivalent to the Warsaw Salem are seen throughout the mid-west and southeast.

The easternmost section of Tennessee, ten to fifteen miles in width, is a true mountain zone with high rugged peaks and dense forests of the southern Appalachians. The flora and fauna of Great Smoky National Park is quite different from that of states to the north and west. East of the mountains is the Valley and Ridge province, followed by the Cumberland Plateau and mountains with the escarpment revealing many miles of Mississippian strata.

Celestine inclusions in geodes are found in Indiana, Kentucky, and Tennessee.
(PHOTO BY LARRY DEAN, WIU; DAVID HESS COLLECTION)

Coral head geodes have elaborate botryoidal interiors. 5 inches.
(PHOTO BY DAVID PHELPS; ESTES COLLECTION)

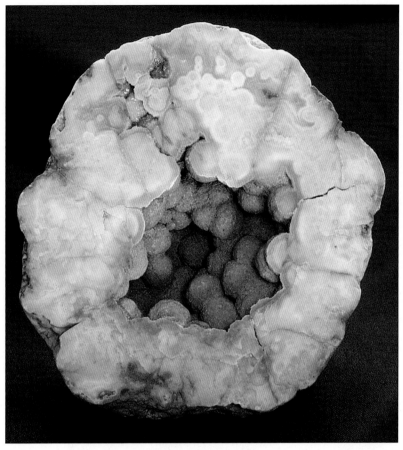

The Highland Rim, the largest of the eight sections of the state, is underlain with limestone of the Mississippian. Tennessee has a series of well-planned state parks, encompassing far over a million acres. Fossils and geodes are abundant in many spots.

Cannon County, southeast of Nashville, is noted for its quartz minerals, mostly geodes. A well-known site is near Woodbury. Tributaries of Stones River are popular locations too. These geodes are close cousins of the Keokuk geodes of the Warsaw formations. The rough wart-like exteriors have caused local experts to theorize as to their fossiliferous family tree. Many of these geodes are reminiscent of coral heads.

Weathering out of the limestone with numerous marine fossils, some of the geodes resemble brachiopods and crinoids. Many of the large sized ones are typical of what has been called "puffed geodes," so blown up that they have little or no clue to their growth in ancient seas. Ranging in size from marbles to basketballs, the geodes have bright quartz linings with inclusions of goethite, fluorite, calcite and pyrite to name a few.

Some are solid, some are thin and fragile, and some lined with tinted botryoidal chalcedony or common opal instead of crystals. Light blue celestite and pink dolomite are considered prize inclusions. Although tons of these geodes have found their way into collections of professional and amateur mineralogists, local expert Virgil Owen estimates that there are many tons left.

Lawrence County geodes are similar, except somewhat smaller. They are found in road cuts, stream banks and in the area of David Crockett State Park. Geodes are also numerous in Clay County east of Oakley. Another location for geodes and agates, too, is Horse Mountain near Shelbyville. Some of the agates, as well as the geodes, have an outer appearance resembling corals, particularly those with honeycomb-like patterns.

Fentress County and its neighbors Overton and Warren, are in the celestite district of Tennessee. Celestite deposits are found in Mississippian limestones

This 3½-inch geode found near Knoxville, Tennessee, has an iron stain, common in this area. (PHOTO BY DAVID PHELPS; ZEITNER SPECIMEN)

There are numerous sites for Tennessee geodes, among them Oak Ridge and around the TVA lakes.

through large parts of these counties. Some of the excellent celestite crystals occur in the quartz geodes of the district with associated inclusions of calcite, dolomite, marcasite, pyrite and others. Some of the locations are near Buffalo Cove, Allons, Livingston, Pilot Knob, and McMinnville.

One time years ago when Albert and I were staying near Knoxville, we were told that small geodes have previously been found on the spreading campus of the university. We were given permission to look and we did find a few small quite ordinary quartz geodes.

Greene County has super sized geodes with rough grayish exteriors and shining linings of water clear quartz. This county also has some thin-shelled geodes. Some of these host loose crystals, often double-terminated, that rattle around in Mother Nature's own rattleboxes. These are from dolomitic limestone called Chepultepec. Geodes are found near Chestnut Ridge and around Midway and Mosheim.

Geodes have been found in riverbanks and around the lakes of the Tennessee Valley Authority, also known as the TVA. With the Mississippi River as its western boundary and the Appalachians highest mountains to the east, Tennessee is a state of enormous variety.

Years later, on another visit to Tennessee, I found some rather small geodes near Oak Ridge. They approached roundness as near as any geodes and had smoother skins than most. The reason I kept quite a few was the unusual exterior coloration, which included most tints and shades or red orange. They were so admired by family and friends, I doubt if I have a single one left.

Kentucky and Tennessee are mentioned most frequently in papers about Keokuk type geodes and geodized fossils, but some papers also include Alabama and Georgia. At a Tucson show I saw a nice geode from Alabama. The label did not have the county name. Nevertheless, I asked the dealer if he had another one, and he did. It is definitely a bivalve and looks much like my South Carolina specimen with a hard gray coating on top and light yellow calcite crystals inside. Two internal fractures had been healed with bright white chalcedony. The edges are coated with what appears like sediment coated with dark chalcedony. It measures 3 inches x 4.25 inches. I think it was broken open with small tools.

A Russian friend told me Russia, Ukraine and several other countries have geodized fossils. Later I was able to examine one from Ukraine. The American deposits may be just the tip of the iceberg!

I am sure there are many more sites in the Southeast. Once years ago when Albert and I were traveling through South Carolina, we noticed a place along the road where the state had mined a shell pit for road material. We stopped as soon as we could find a place to park and walked back to the big pile. Most of the broken shells were in poor condition, but I did find one keeper. It was protruding from the far edge of the pit about an arm's length down. It was a complete bivalve. Most of the top and bottom, instead of being dirty white like most of the pile, was gray. Sticky sand was clinging to the outer edges. Albert took out his pocketknife and started cleaning the edges, and the thin shell fell open to reveal a wonderful lining of pastel yellow transparent calcite crystals. The exact classification of calcite crystal lined shells is unclear.

These beautiful mysterious geodes are sometimes called *clam geodes* or *shell geodes*. They seem to be found in fossil shell deposits in most of the East Coast states, much to my surprise.

Clam geodes from Virginia Beach, Virginia, have white or clear calcite crystals. Some from near Westover, North Carolina have opaque creamy white crystals. A shell pit near Chestertown, Maryland, revealed several clam geodes with vivid yellow calcite. Since the fossil shells are mined for road material most are broken. If I had not found that fine example near Myrtle Beach on a day the machinery was not working, I might have never believed in them.

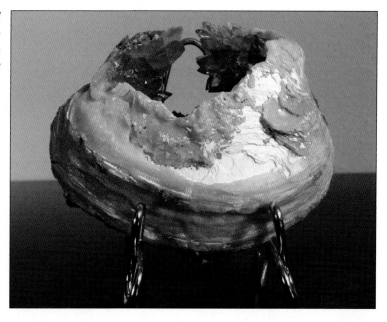

A partially silicified clamshell from the Georgia/Florida border is lined with gold colored calcite crystals. 3¾ inches. (PHOTO BY DAVID PHELPS; ZEITNER SPECIMEN)

I just heard that some pieces of fossil ivory near Saltville, Virginia, are lined with delicate calcite crystals. Now that I have to see!

We had found many natural wonders during our thirty years of travel to the rock, gem, and mineral sites of North America, so that find became eclipsed by time, until the day I was researching an excellent article about geodes that mentioned geodized fossils of the Southeast. I suddenly jumped up and ran to the basement saying to myself, "I've got one! I've got one!" I took the still beautiful South Carolina geode out of a fossil case.

Now I am wondering about some of the unusual geodes of north Florida!

Mineralogist Tom Loomis found this geode on Nemo Road a short distance from my house. 3½ inches. (Photo by David Phelps; Zeitner Collection)

CHAPTER ELEVEN

Amethyst Surprises and Baby Rattle Rocks

— JUNE CULP ZEITNER —

At a recent show in South Dakota I was astonished at the beauty of a case of unique amethyst geodes, and even more surprised when I found they were from South Dakota's mountain range called the Black Hills. I rate them as the best amethyst geodes in the United States, and probably the best of any sedimentary amethyst geodes.

The display, by Joe Nonnast, included over thirty flawless, highly polished geodes. Joe, a long time gem, mineral and artifact collecter, said they came from a rugged and remote area, and out of a big pailful only two or three were worth polishing. I asked Joe if David Phelps, one of the photographers for this book, could come with me to his home to get some pictures and get more details about the geodes. Later, when we arrived, Joe took us to his basement and showed us his incredible collection of geodes, agates, and agatized wood.

The geodes range from about 2½ inches to 6¼ inches in size. The exterior surfaces are botryoidal and the general shape more oval than round. The outer limestone coating is gray stained with yellows and reds. The inner porcelain-like shell enclosing the amethyst center often has a very pale blue border shaped like fluffy summer clouds. This runs into a snowy white to light cream shell with scalloped edges outlining the amethyst center. This light area contrasting with the amethyst is from ¼-inch to 1¼-inch wide. The amethyst varies from lavender to medium purple. The purple colors are enhanced by the contrasting pastels.

*South Dakota amethyst geodes are among the most beautiful
gem quality sedimentary geodes. 4 inches.*
(PHOTO BY DAVID PHELPS; JOE NONNAST COLLECTION)

The colors are true and clear, not a bit muddy. While crystal shapes arc readily apparent, the crystals are packed so closely together that there are not many openings, so some centers are solidly full of amethyst. Rarely, a pastel peach or warm buff colored pattern enhances the white border. The border surrounding the amethyst is wide, opaque, and resembles fine porcelain.

Some of the shadowy patterns in the geodes are mysteries, but at least one example shows a distinct *fossil brachiopod*. Some of the knobby exteriors have attached segments, cemented and healed by nature. If the rounded knobby exterior is measured, it is surprising how much bigger the circumference is than it appears. We went to take some pictures of the geode-bearing formation, a distinctive dolomitic limestone called the Englewood. The Englewood can be seen near Piedmont and on Squaw Creek, and near some old railroad trails in Lawrence County. Other outcrops are near Deadwood and Spearfish.

The primeval Englewood Formation has been variously described as pink, reddish, rusty, purplish, terra cotta, and russet, but actually an artist will see all these colors, plus a healthy dollop of ochre. Topped by the Pahasapa Formation, both are described as fossiliferous. Other inclusions in the geodes are ankerite, siderite, selenite, limonite, and hematite although none of these is common. Willard Roberts reports rutile as an esthetic inclusion in some Lawrence County amethyst. Rutile is also present in the Red Feather Lake's amethyst of southern Colorado.

The geodes are solidly embedded in the hard limestone and must be worked out by hand with extreme care. Sometimes there are several close together, but most have few neighbors. In the cliffs there are also occasional vugs lined with small bright quartz crystals.

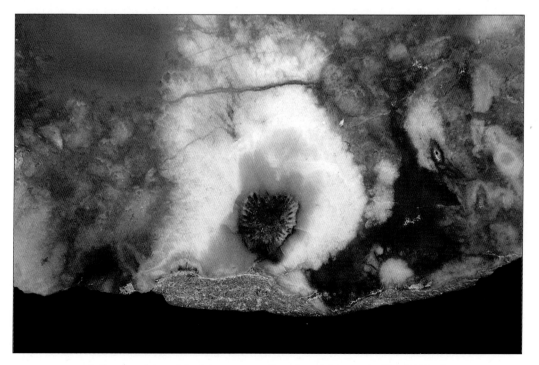

A tiny brachiopod is at home near the edge of this geode. Fossil ¼ inch.
(PHOTO BY DAVID PHELPS; JOE NONNAST COLLECTION)

Some of the geodes have no amethyst, but are lined with sharp clear calcite crystals. Such geodes have more of the peach color and more pattern in the outer portion. The igneous uplift of the Black Hills pushed aside a multitude of ancient vast sedimentary formations left by Cretaceous seas, which once covered the region. More of the igneous uplift is exposed in the Harney Range of the southern Hills.

A short distance from eastern Wyoming, Lawrence County, South Dakota is a scenic mountain area mostly covered by Ponderosa pines and aspen with some spruce, oak, birch, willow, juniper, and chokecherry. The area was prospected for gold in the late 1800s. The Homestake, long the nation's most productive gold mine, is not far from several Englewood exposures and the mine itself has yielded fine amethyst crystals (Roberts). The Englewood Formation can be seen in some of the old open cut mines of the region. Another site in Lawrence County yields oblong agates and geodes with dominant colors blue and russet instead of violet and peach.

In some of the geodes with the thick white porcelain-like shell there are inclusions of what appear to be fluffy clouds of a faint creamy peach tint encircling the upper portion. This seems to enhance the depth of violet in the amethyst.

Joe doesn't know if others have collections similar to his, but he has seen places on Deadwood Hill and in Spearfish Canyon's sheer cliffs where there has been digging in the Englewood. Broken pieces he has found in rockslides reveal more white and water-clear quartz, but no amethyst.

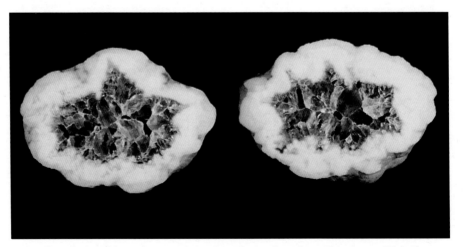

Scalloped edges are a characteristic of Englewood geodes. 2½ inches × 3¾ inches.
(PHOTO BY DAVID PHELPS; JOE NONNAST COLLECTION)

154

Blue agate geodes are a variation of Englewood formation treasures. 4 inches.
(PHOTO BY DAVID PHELPS; JOE NONNAST COLLECTION)

In *Minerals of the Black Hills* Roberts and Rapp mention amethyst geodes on Whitewood Creek. Years ago my husband and I hunted along the creek and found a few disappointing geodes with no resemblance to Joe's collection. South Dakota pioneer mineralogist Samuel Scott noted the occurrence of amethyst geodes along Bear Butte Creek in Lawrence County.

The best geodes may occur in the upper part of the Englewood, which has been eroded away in many areas. This formation was once assigned to the Devonian, but it is plainly a Mississippian formation which has been productive of amethyst geodes. There are numerous Englewood exposures in the northern Black Hills, but very few in the southern Hills, where they have been spotted in a few old mines. The best known exposure of the Englewood is in Spearfish Canyon, and was plainly seen in the movie *Dances With Wolves* which swept the Academy Awards, and helped make Spearfish Canyon a "must see" destination for the millions of tourists who annually visit the Black Hills.

The Bear Lodge Mountains of Wyoming adjoin the northern Black Hills near Spearfish. Upper Cambrian rocks are exposed in Spearfish Canyon. Nearby ski areas are mostly Mississippian. Limestones are cut by intrusions of Tertiary rhyolite. The oldest known rocks in the Black Hills are the exposures of Little Elk Creek Granite on Forest Road 135 near Dalton Lake. This granite gneiss is about 2.55 billion years old. Throughout the northern Black Hills old open pit mines reveal many sedimentary formations worth additional investigation.

LAPIDARY TREATMENT

Dark amethyst is also found in centers of Tepee Canyon agates west of Custer. These agates are nodules occurring in sedimentary rock, not geodes, although the crystals in vugs in the agates are sometimes dark amethyst. Lapidary treatment of the amethyst geodes is not easy. The pale bluish edging around the white or creamy white section is softer than the pastel area, and the deep basically white border is softer than the amethyst. Undercutting and pitting may be hard to avoid. (Are there clay inclusions in the chalcedony?)

Asked how he gets such professional results, Joe says he uses diamond as well as his many years of lapidary experience. He cuts the geodes in half. Noting the considerable hardness differences, he carefully grinds and sands the surfaces. After the pre-polish with 14,000-grit diamond he gets a superb polish with cerium oxide.

Joe feels that the geode halves polished for permanent display take a glossier polish and look more attractive if slightly rounded rather than flat. If only half of the geode is high quality, it is carefully rounded for display. In addition to the virtually flawless geodes, Joe has found wide vein agate with approximately the same colors, only the lavender is replaced by blue, and there is more of the peach grading into warm buff. Brought up on a farm near Whitewood, he has spent countless hours exploring remote parts of the northern Black Hills.

The Englewood geodes are definitely top quality and are far different from South Dakota's better known contribution to geode lore – the baby "rattle rocks" of Imlay. The Englewood geodes are gems; the rattle rocks are fascinating and fun.

Some of the geodes, although still having unusually thick shells for their size, have bigger hollows than most of the amethyst ones. Many of these hollows are lined with calcite instead of quartz. A frown shaped cavity of a cut geode lined with white hexagonal dogtooth calcite crystals has an angry look. The pale clouds near the outer part of the shell are identical to those of the amethyst examples.

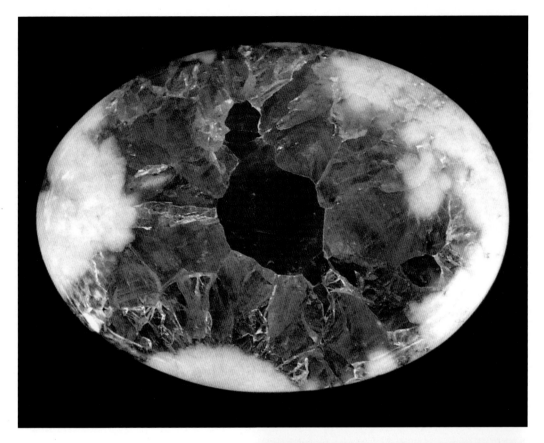

Above: and Right:
Joe Nonnast cut this 30 × 40 mm cabochon
from an Englewood amethyst geode.
He also made the pendant.
(PHOTO BY DAVID PHELPS)

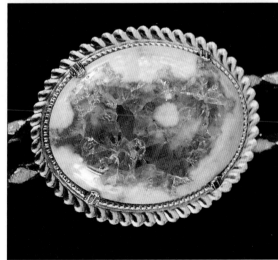

It isn't often that a book gets to describe a geode that has never been described previously. However, these South Dakota amethyst geodes although mentioned in a few words in both Scott and Roberts are not described by them or anyone else as far as I could find. I searched through back issues of *The Mineralogist, Earth Science, Lapidary Journal, Gems and Minerals*, and *Rock and Gem* and found nothing. Perhaps this is because South Dakota's Fairburn agates and abundant rose quartz as well as cycads and dinosaurs have attracted so much attention, or maybe the geodes are so little known, even here in South Dakota, that they have not inspired gem and mineral writers.

There are certainly fewer of these than most of our state's mineral and lapidary treasures, but many rare and fine American lapidary materials are not widely known, for instance the emerald green translucent williamsite. So here, at last, is the formal introduction of the Englewood geodes.

Formula	SiO_2
Hardness	6 – 7
Density	2.65
Varieties	Cryptocrystalline, crystalline, and amethyst quartz
Sizes	2½ in. to 6¼ in. diameter
Colors	White, pale peach, violet
Crystal System	Hexagonal
Source of amethyst color	Ferric iron, and/or manganese and titanium
Source of peach color	Ferric iron
Mode of Formation	Low temperature and pressure
Luster	Vitreous
Location	Englewood Formation of South Dakota's northern Black Hills
Collector and lapidary	Joe Nonnast

BABY RATTLE ROCKS

Like their Black Hills cousins, the chalcedony "rattle rocks" of the badlands areas of western Pennington County occur in sedimentary formations. Often referred to as baby geodes or baby rattle rocks these geodes are mostly two inches, more or less, in diameter. A really big one is 4 or 4½ inches. The rattle is because fragile loose crystals or tiny balls of wee quartz crystals have broken off. It's frustrating to decide whether to keep them as "rattlers," or to open them and see the precise cause of the rattle.

The classic site for thousands and thousands of these geodes is near the ghost town of Imlay on Highway 44 and surrounding ranch lands. We found years ago that the best beds were where the railroad track was closest to the highway and a little farther south. This land is now part of the Buffalo Gap National Grasslands but there is some private land here too. The biggest creek in the area

Chalcedony quartz geodes sometimes rattle because there are loose crystals inside. 2¾ inches.
(PHOTO BY DAVID PHELPS; ZEITNER SPECIMEN)

is Cain Creek, which has branches on both sides of Imlay. The area is hilly but not steep or rough.

I have several geodes in the four to five-inch size and a few in the smallest size, about one-half inch across. Many of the geodes are really round. Most have thin shells, usually white, or off-white. Albert Zeitner found an area quite a distance from the railroad tracks with lots of small pink geodes. Last time I visited the locality was to see if I could find enough pink ones, one inch or less, to tumble and drill for beads. I found enough, but never got around to the project because the thin shells are very rough and irregular and present quite a challenge.

Albert found another place where the geodes were dotted with red. My late father-in-law, G.B. Zeitner, found a bed of bright orange geodes in this region back in the early 1930s. He collected all that were exposed and inlaid them in a concrete birdbath for his garden. When we sold the property years later we couldn't find a way to remove the birdbath without tearing up that part of the lawn. So we left the orange chalcedony birdbath, a decision I deeply regret.

The orange geodes were mostly flattish, about three inches in diameter with shells a little thicker than white ones. The orange color was intense. Other parts of the early Zeitner geode collection, including broken halves, large whites, small pinks, lavenders and spotted examples, were inlaid in a low concrete border around a park area in the center of a tourist cabin camp my father-in-law owned. The last of the orange geodes may still be there – the broken halves only revealing showy cavities.

Like my husband, my father-in-law covered many miles in a day of rock hunting, but he told me the orange geodes were near Cain Creek west and a little south of the old railroad tracks. Except for some deeded property, that region also is now part of the Buffalo Gap National Grasslands.

I don't know when the Imlay geodes were first found, or who discovered them, but a 1919 book by C.C. O'Harra, published by the South Dakota School of Mines, notes the "geodes of the White River badlands." "The prettiest ones (geodes) of rather small size are found near Imlay. They commonly have an irregular shell of chalcedony, more or less filled with bright clear-cut white or colorless quartz crystals varying from microscopic size to one-half inch or more in size."

O'Harra also reported those with smaller crystals were called "sugar geodes," a name which has not persisted. O'Harra describes the many veins of chalcedony in the area and says the geodes are undoubtedly connected with the abundant chalcedony deposits. Some of the chalcedony is blue and of cutting quality. Some is pink, sandwiched between two layers of blue, or blue between the pink. There

Geodes occur in badlands areas of southwestern South Dakota

is also white and red, black and red and black and white. These "grasslands" were first explored by scientists in the late nineteenth and early twentieth centuries.

It used to be easy to walk along the tracks and gather a big sack full in a few hours. The ground is hard and arid, for the location is not far from the White River badlands, a huge area where many varieties of quartz have been found for over 150 years. One has to walk quite a ways from the tracks now to have good hunting. Much of the land is national grasslands, but there is also deeded land, and even the public lands may have restrictions. In the past, several clubs from South Dakota and neighboring states have had field trips here.

Some of the little geodes are grown together, making specimens that appear at first glance to be huge geodes for this area. I have one with two 2½-inch geodes grown together and have seen a number of similar ones. A specimen I respectfully call the "mother" geode has six elaborately shaped geodes grown together as if they were perfectly fitting parts of a master's intarsia. We cut this

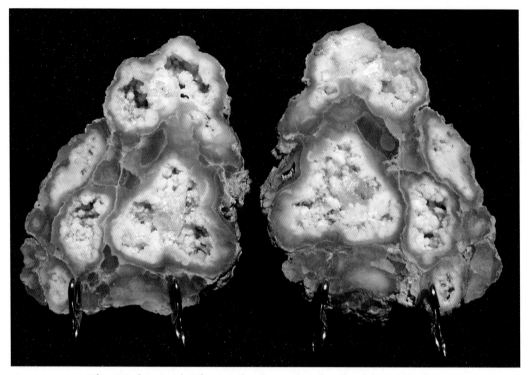

The "Mother Geode" from Imlay has a close family of six. 5¼ inches.
(PHOTO BY DAVID PHELPS; ZEITNER COLLECTION)

lengthwise to examine the cavities and the inclusions. I don't know of another geode with so many cavities coming from this bed, but I am sure there must be more. Whether such geodes have two cavities or six, I call them multi-cavity geodes. Most of this type of geode has been found on privately owned farmland near Imlay, not on the classic site on the public land.

We became aware of these "multis" when we were hunting along the railroad track and a group of children walked by. A little boy came running towards us and said, "Bet I know what you are picking up. It's geodes. But I have some in my yard that are lots bigger." After some friendly conversation the boy led us to his parents' ranch only a few hundred yards away. He was right. They were bigger because nature had glued them together!

The multi-cavity geodes must have been small geodes growing simultaneously in crowded circumstances so touching each other they were shaped by those that they touched and grew together becoming inseparable. Cutting a multi-cavity geode reveals that each cavity is one geode with its own chalcedony shell, and its own crystal lining. Each part of the apartment complex has the same colors and inclusions but with different arrangements and emphasis. Each part has a warm light tan skin separating it from its neighbors, but also grown to the skins of the neighbors. Each part has an encompassing blue gray chalcedony shell lined or partially filled with minute white, pastel, or colorless quartz crystals with the same inclusions as the other parts.

Studying the exterior of the multi-cavity geode reveals extraordinarily rough tan skins grown together, however in spots the tan skin has chipped off, revealing the blue gray chalcedony. To me this is a problem. Why would the skin chip off so crisply and cleanly? It is hard, difficult to scratch, but is thin as onion-skin and as brittle as a piece of old glass. But the color of these skins is remarkably like the color of the gumbo soil where the geodes are found. All this leads one's theories in an entirely different and even more puzzling direction.

Vince Henderson found a pure white Englewood geode just around the corner from my home. 2½ inches.
(PHOTO BY DAVID PHELPS; ZEITNER COLLECTION)

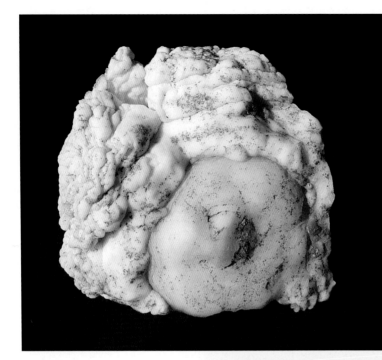

When broken, this geode revealed a unique interior. Back, 3½ inches. (Photo by David Phelps)

The inside of the geode resembles a cave after an earthquake. 3 inches. (Photo by David Phelps)

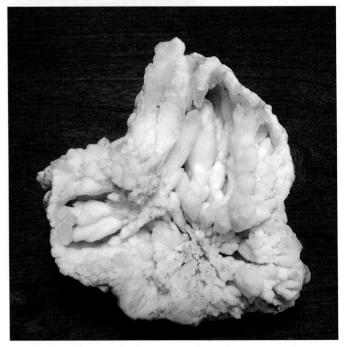

Similar geodes to the Imlay products are found in Badlands National Park. Sparsely scattered over wide areas, they have not been collected or studied to any great extent. Partially because no collecting is allowed except to Universities with permission – (usually granted for just certain vertebrate fossils). And secondly because even before the National Park was established few people collected rocks in the "Big" badlands because the area is remote, rugged, steep, and dangerous.

Inclusions in the Imlay geodes are usually rare. Some of the rattling crystals are double terminated. The drusy quartz crystal linings are most often snow white or crystal clear, but pastel pink or lavender linings are desirable enough that finders keep opening these little boxes as if they held rare treasures. Occasionally a geode will have a pale citrine or smoky quartz lining.

The most common inclusion in the Imlay type geodes outside of some member of the quartz family is selenite. Like most very small crystals the little selenites are astonishingly perfect and beautiful. The shining monoclinic transparent crystals are thin single blades, divergent clusters, or rosette groups. The crystals may be liberally scattered over the interior cavity lining or may be confined to one small part of the cavity. Tiny bladed crystals of selenite are amazingly brilliant, especially when viewed with magnification and good light. My own candidate for finest selenite inclusion is a small group centered in a good cavity lined with white drusy quartz. The selenite cluster resembles a miniature Mayflower-type of sailing ship ready to leave for a foreign shore. Pastel pink, yellow, and peach calcite crystals are rare inclusions. A not so common inclusion is marcasite. A few geodes have sported black acicular crystals of goethite.

The geodes of this area are seldom solid. About fifteen percent of them are rattle rocks, and many are so light that it is evident they are hollow. Before the ones that rattle because of loose crystals were called rattle rocks, they were called rattle stones. The lost name "sugar geodes" must have been referring to the granulated sugar sparkle of the infinitesimal crystals.

The geode that gave me the most excitement when I opened it yielded an incredibly tiny perfect sphere of perfect ice-clear quartz crystals. Each gleaming crystal was the same size and each pointed outwardly at the correct angle. This quarter-inch quartz crystal sphere is the smallest item in my best thumbnail box. I suppose it should be in with my micromounts.

Chalcedony/quartz geodes similar to those of the Imlay area have also been found along Medicine Creek which like Cain Creek flows into the White River. Medicine Creek is in the Pine Ridge Sioux Indian Reservation farther south than the better known Imlay and Cain Creek locales. White River is a major river in western South Dakota. The immense badlands are named the White River Badlands. Geodes with thin chalcedony shells have been found on Bouquet Table, Heck Table and other table lands, and surroundings south and east of the town of Scenic and all the way to Conata, another railroad ghost town.

I have known about the Imlay chalcedony geodes for about sixty years, but I was totally unprepared to have South Dakota make geode history with the magnificent geodes of the Englewood Formation.

CHAPTER TWELVE

Florida's Ocean Harvest

— JUNE CULP ZEITNER —

The agatized coral geodes of Florida defy definitions of geodes. Strange and complicated in shape, seldom having one crystal lined cavity, as most sedimentary geodes do, and with great variations in color and size, these geodes rank among the most unique in the world.

Several locations for these extraordinary specimens have been found in other parts of Florida and even Georgia, but the classic location and initial discovery was in Tampa on Florida's west coast. When was the remarkable agatized coral reef discovered? No one knows for certain. Well-shaped spear points of chalcedony geode shells have been found. Undoubtedly the first settlers in the Tampa area dug up geodes when they were building dwellings and roads, but found these disgusting muddy objects to be worthless nuisances. The first mention of these geodes in print was about 180 years ago, shortly after Florida was purchased from Spain. The mention attracted little attention and it was not until James Dana's third edition of *System of Mineralogy* in 1950 that Tampa's "chalcedony, carnelian, agate, silicified shells and corals" brought more notice.

Gradually Ballast Point, which juts out into Hillsboro Bay about three-quarters of a mile, became famous in the gem and mineral community. Previously Ballast Point's only claim to fame was a Civil War battle.

The Tampa Formation, a one-hundred-foot layer of limestone, was laid down during the Miocene. The formation can be traced for sixty miles north and south of Tampa and forty miles east and west, with the western extremity on the Gulf of Mexico. Since the corals show little sign of abrasion, it is probable that few were ever carried very far from the original reef.

Good blues are not common in sedimentary geodes except
those from Florida. 4 inches.
(PHOTO BY DAVID PHELPS; DAVID PHELPS COLLECTION)

It is said that in the mid-nineteenth century early residents with salesman-ship talents sold the corals they found to newcomers. This was shortly after T. A. Conrad wrote about the coral geodes in an article in the *American Journal of Science*.

The agatized corals of Tampa Bay, actually Hillsboro Bay, may be the world's most unique geodes, or at least tied with the angular geometric polyhedroids of

Brazil. The extensive coral reef stretching far out into the Bay, as well as just under the surface of the land, was a living wonderland, an undersea Eden over 20 million years ago. Then something disastrous happened. The reef began to die. However the tragedy of the reef became a bonanza for agate and geode collectors, scientists and artists.

A process called metasomatic replacement is responsible for transforming the calcium carbonate reef into gem chalcedony, a variety of quartz, SiO_2. As the calcite passed into solution through capillary action, almost simultaneously the calcium was replaced by silica.

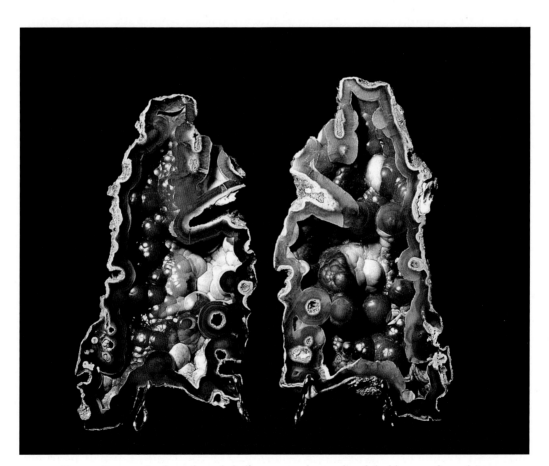

Tampa Bay agatized corals and shells are pseudomorphs of chaldecony after calcite.
4 inches × 7 inches. (ZEITNER COLLECTION)

This reef included many species of coral, plus a variety of shellfish, worms and barnacles, similar to those found in warm coral reefs today. It wasn't until the early 1930s that the agatized corals and shells became collectables. Since then many tons of this ancient reef have been collected and studied by scientists and amateurs, cut by lapidaries, and displayed in museums around the world.

The corals are pseudomorphs, but not the usual kind. The coral origin can be clearly seen. The species is often identifiable, but seldom is much of the original coral preserved. The gemmy interior is new and different. Polyps are missing, boring clams have altered the appearance. Fingers have fallen off and become separate small geodes. Barnacle guests have become silicified. Solid corals are rare here. Most of these geodes were colony corals, and most are now hollow.

The waters of the Miocene seas of the Tertiary period when the reef was flourishing were clear and warm. What happened? Was there a sudden drop in temperature? Was there pollution? A disease? We can't tell for sure, although for now the temperature change from warm to cool is the leading theory. By far most of the Tampa geodes come from a peninsula called Ballast Point. The name came from an old idea that the geodes were ballast brought over by ships from Europe long ago. The ships were then loaded with Florida products for return trips. For years, people believed these strange rocks could not be of local origin. Acceptance became general after Davis Island, a man-made island, was created. Deep dredging brought up load after load of rocky soil, and the "rocks" were all coral geodes from the large ancient reef.

Willard Olson of New Port Richey, Florida took us to our first trip to Ballast Point in 1950. The point was then a city park. There was an old public restroom there and little else. The air smelled like rotten eggs. There were strangely-garbed individuals knee deep in mud as far as one could see. Some wore rubber waders and were working in the water with pry bars, others were wearing rain coats and were sitting in the mud beside the glory hole they had dug, which was rapidly filling with water. Several wore bathing suits, galoshes and sun hats. It was hot and humid, but the tide was low and everyone looked happy. Cautiously I approached a few people to see what they were finding. I was not impressed! What they were smiling about was pails full of the ugliest rocks I had ever seen. I had been an avid gem and mineral collector since the 1930s, but half-heartedly gave this a try, only because Albert was already feeling around in the muck with his Estwing.

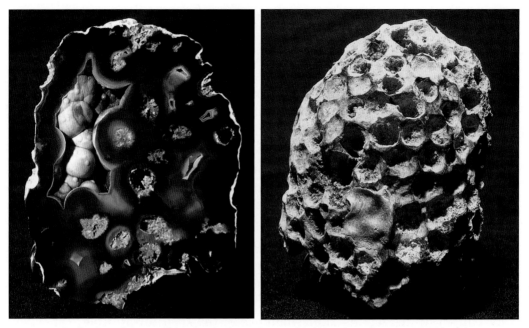

The back and front of this geode show the coral origin more than most. 2¾ inches.
(PHOTO BY DAVID PHELPS; ZEITNER SPECIMEN)

Noting my lack of enthusiasm, Willard invited us to come to his home to see his collection. There I was soon out of "ohs and ahs and wows!" because what I was looking at was the most interesting, most exciting, and most esthetic geodes ever. They were cut and polished. The shapes were unlike any kind of geode or gem material I had ever seen. The colors glowed. The drusy quartz sparkled. There were smooth grapelike botryoidal cavities, elaborate and complex. The color spectrum included red-orange (carnelian), orange-brown (sard), black "onyx", and yellow, peach, cream, blue, brown, sepia, ivory and white. It was early fall, but the days were still long. I hope I thanked Willard profusely enough before I said to Albert, "I think we could go back to Ballast Point for a few hours before dark."

We returned to Ballast Point many times after that during several years of discovering that those who had told us there was nothing to collect in Florida had it all wrong! We saw some of the arrowheads, spear points and other coral items the Caloosa Indians had made from agatized coral. Dean and Rae Thompson had some points up to five inches in length. Archeologists theorize that carefully shaped geodes found in Caloosa sites may have been used as utensils.

We found that some of the geodes fluoresced and a few phosphoresced. We learned that luck was slightly better by working in the water at neap tide with a long metal pry bar. Sometimes we would strike an old bottle, a broken ceramic jar, or a steel can full of clay, but often we would come up with a bottle shaped, fan shaped or hour glass shaped geode. I never again defined geodes as round.

Several kinds of geodes became favorites of mine. One type was finger geodes a little over an inch in diameter which could easily be slabbed into freeform agate earring hoops. Another was the geodes which had several colors, preferably blue and black, covered with shimmering drusy quartz. The third was high on my list because it was the perfect example of the wonder and mystery of geodes. This kind of geode is called an enhydro. Picture an artistically shaped transparent sealed vase of water, and you will have the image of a clear chalcedony geode with water rushing from one end to the other as you move it.

We were out on the beach one Sunday morning specifically looking for enhydros. Every time we found a transparent colorless geode we would hold it up toward the light and move it upside down. A group of three ladies, dressed as if for church, were walking along the edge of the road toward their car when they spotted us. Evidently they were curious about our unusual behavior, but also suspicious that there was something seriously wrong with us. Curiosity won. One lady, wearing white gloves, customary at that time, cautiously approached me to ask what I was looking at. I held up my best find and let her see the quarter of a cup or so of water race back and forth in its chalcedony cage. With white-gloved hand she reached and asked if she could hold it and tip it back and forth. The entranced ladies left us about an hour later.

The best geodes are usually dug from clay full of gritty inclusions. After cleaning the surface as much as possible, one can see rough cherty surfaces with distinct patterns of star-like or honeycomb-like coral. The geodes are often further ornamented with a shiny piece or two of mother-of-pearl from a shell, or a wee barnacle, also victims of the death of the reef. Often the next layer is chalcedony or agate, which may be less than a half-inch thick or over an inch. The third, or inner, layer is the surprise and/or prize. It may be a velvety botryoidal layer, beautifully patterned, colored, and textured, which needs no lapidary treatment ever. Or it may have stunning colors coated by what appears to be diamonds. A drusy layer can be the frosting on the cake. Geodes, when cut in two, may reveal one or more small round holes made by burrowing clams or sea worms. But the holes are bordered by agate and are an extra attraction.

In some well-known localities, geodes of a specific area tend to resemble each other in some way – size, color, or inclusions, – but the Ballast Point geodes, even a group dug from the same hole, vary widely. The exception is that the Tampa residents, plagued by these geodes when they are trying to plant trees or gardens, find a majority of red ones. Perhaps being on land for a few centuries allowed iron in the soil to color them. Tampa geodes, whether from land or water, can easily be distinguished from Florida geodes from other localities such as the Suwannee River or the Cross-Florida Canal.

Several serious collectors, some thirty years ago, invented what they called a "Coral Finder." Whether it really helped a lot or whether the men were just lucky is not known, but they did find some dandies. The device was a stainless steel rod, well sharpened at one end and with a comfortable handle at the other. With a little practice the user was supposed to know when the gritty coral bearing layer had been reached.

Rarely, a geode is found with two individual, but connected, cavities. Usually such geodes have distinct differences in color and texture; for example one of mine has a blue botryoidal lined cavity and a yellow chalcedony lining frosted with clear transparent crystals of drusy quartz. Another has a blue lining with a black shell, and its smaller connected cavity has a black shell with a cream and brown variegated botryoidal cavity. The strangest is a pair of Siamese twins. Both parts are the same size and share a continuous half-inch shell of clear inward facing quartz crystals. However, in one half there is a ball of outward facing quartz crystals touching the inward facing crystals, and right in the center of the ball is a small hole which on examination shows that the hole was made by a burrowing clam. Another clam is stuck on the outside, but too small to have made the hole. Also on the outside of this puzzling specimen are several small thin fortification patterns.

In the case of the dual-chambered specimens with quartz crystals in one part and chalcedony in the other, E.H. Lund wrote in 1960 that both must have formed under the same temperature and pressure. A small almost invisible hole on the exterior of the chalcedony side allowed water to go in and out, but there was no vent on the quartz side, which had sealed itself. He also said in his *American Mineralogist* article that the crystal side, like many solitary crystal geodes, was an enhydro.

W.H. Dall published a monograph in 1915 about the Tampa deposit including the coral species he had identified and studied, and the agatized mollusks and other fossils which were still calcium carbonate, or only partially silicified. He

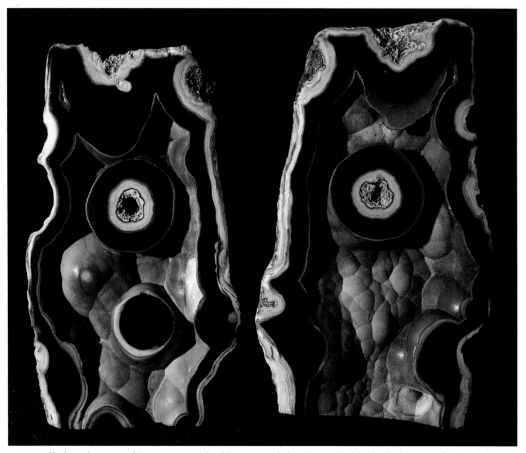

A small clam burrowed its way into the living coral, leaving a hole which became the focal point of a Tampa Bay geode. 4½ inches. (Photo by David Phelps; Zeitner Specimen)

described 312 mollusks. He thought the bay water was silicifying the shells, perhaps more rapidly than earlier scientists had postulated. (Ballast Point at the time of Dall's study was the terminus of the Tampa streetcar line.) Dall's monograph far exceeded an 1886 survey that listed only 47 mollusks. At the time of Dall's work, several feet of the Tampa limestone formation were exposed along the water and the agate geodes were plainly seen. Now there are concrete and rock walls along the shore mostly studded with "No Trespassing" signs.

Most of the geodes are taller than they are wide. Many are in the range of four inches to seven inches high. Some "hand" shaped or "fingered" ones are about five inches across and six inches high. The record in size for years was a 220-pound almost complete coral head found in 1961 by Arthur and Olive Breu of Saginaw, Michigan. Sixty inches by sixty-six inches in circumference, a small hole in the side reveals an elegant and multicolored interior. I saw this monster at a Midwest Federation show. The coral pattern was unmistakable. Olive joked that it was probably a brain coral. She said it was attached to a reef five feet deep in the bay and they had to break it off the little stool of attachment. At the end of the 1960s a true giant was found near the Breu's prize find. Measuring three feet in diameter, it weighed in at slightly over 300 pounds.

I have a specimen so translucent that I can see a long worm track the length of one side. The late Dr. Henry Millson, an authority on the Tampa geodes, found some of the agate worms that made their tracks so long ago. Well-known earth science author Russell MacFall found some of the fossilized boring clams, – one even fits into the coral hole it bored.

Millson reported geodes with iris agate. He also noted the occasional fluorescence and phosphorescence of Florida geodes. Although the colors found in them almost cover the spectrum, no true violet has been found. Sometimes white layers turn out to be cachalong opal instead of chalcedony.

The thin finger corals will float. So will a large thin-shelled one with a big hollow. Inclusions other than some variation of quartz are few. Among these are chalcopyrite, calcite, aragonite, kaolinite, and barite. The deposit extends far under the surface of the land as well as out into Hillsboro Bay.

Almost any excavation reveals some coral. People digging a garden run into geodes. Mac Dill Air Force Base, which has a huge shoreline, is underlain with coral geodes. Agatized coral geodes have been found along Gandy Boulevard, the Clearwater Causeway, Honeymoon Island, and near New Port Richey and Tarpon Springs. Most of the coral near New Port Richey and Dunedin is solid. The Ballast Point area is now closed and strict laws forbid digging in most of the formerly productive areas. Finding geodes is hard, but not impossible.

Fifty years ago Tampa Bay coral was discredited as a jewelry material. Those were still the days when gem material had to be cut in round or oval cabochons and polished to a high gloss. Now, when creative shapes are desirable, and when natural surfaces of tiny drusy crystals or smooth satiny botryoidal surfaces are much admired, old stocks of Tampa material are coveted.

One lapidary back in the 1950s was well known locally for his original agatized coral jewelry. He made spheres, jewelry boxes, bola ties, bracelets, necklaces

A multi-colored geode exhibits a palette of subtle colors. 2½ inches × 6 inches.
(Photo by Douglas Heym; Heym Collection)

and rings, and wrote an article for the *Lapidary Journal* telling how to do it. His name was Dr. O.C. Mehl.

I met Dr. Mehl once and saw some of the inventive and splendid jewelry and decorative objects he made of the Tampa coral. Although he sometimes cut a traditional oval cabochon, he sanded and polished only the edges, the natural botryoidal domed top was then highlighted. Most of his shapes were freeform.

He made bracelets from five freeform slices of agate fingers. Some original silver wire textured bracelets feature only a large exotic botryoidal or drusy piece of exotic coral. His handsome bola ties were also each set with a select Tampa gem. He made small bowls from well-shaped geodes with deeper hollows. He cut off the tops of rounded geodes for little boxes. His "Pandora's Box" was assembled from hundreds of polished squares of agatized coral, almost an intarsia. Dr. Mehl used silicon carbide for grinding, sanding and prepolish. He used tin oxide, cerium oxide or Linde A added to either for the final polish.

Clara Roder had a Coral Museum in Hot Springs, Arkansas. The Roders gathered and cut and polished their own geodes, but enjoyed doing this more when the cut pairs resembled something else. They first noticed this when they cut a large colorful pair which, when opened, resembled a gorgeous tropical butterfly. This led to another pair shaped exactly like fancy valentine hearts. Another

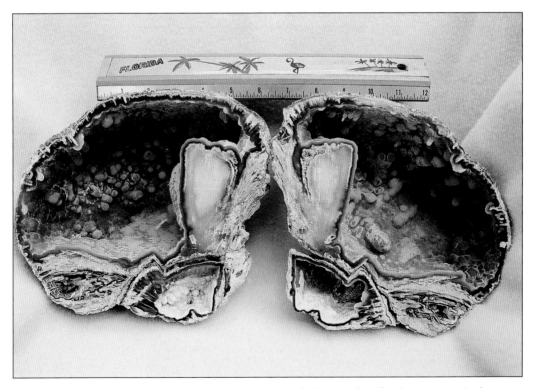

The Heym Collection features aesthetic specimens of Florida's fine fossil gem. 7 × 9 inches.
(PHOTO BY MILLIE HEYM)

pair resembled saguaro cacti, and after that a pair of goldfish. This started their museum, which had hundreds of pairs of high quality Tampa geodes and for years was open to the public free of charge. The Roders are both deceased. I haven't been able to find out what happened to their remarkable collection.

Mostly seen in pairs with polished edge borders, the geodes are adapted to other lapidary techniques as well. Freeform slices of interesting shapes can be used for pendants or earrings after being polished. Tumbled pieces make unusual key chains. A deep half of a carnelian geode makes a nice tray for stamps, rubber bands, business cards or other small items. A more or less oblong geode about as wide as high can have the top cut off to make a jewelry box.

We polished all our geodes with tin oxide after grinding and sanding with silicon carbide 600. We found that some of the geodes were unusually hard. I had a beautiful blue and black piece I wanted to use for a pendant. We had no drill with us while we were traveling. Our dentist in Clearwater saw some of our

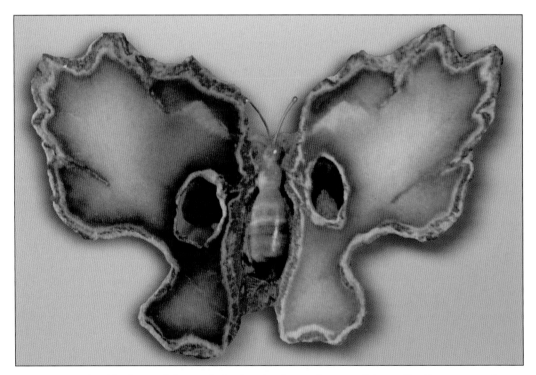

Lapidaries used the corals for jewelry, boxes, trays and decorative items.
Butterfly by Ed Kraushaar. (Photo by Edward Goldberger)

Blue botryoidal chaldedony with dark agate frames are favorites of many Florida collectors. 4 × 4 inches. (PHOTO BY DOUGLAS HEYM; HEYM COLLECTION)

geodes and offered to drill the piece for me. He said it would be no trouble at all. After breaking three drill points he said, "What in the world is this material? I've never seen anything like it." We had it drilled much later by a jeweler who also commented on the hardness.

Perhaps the most fun we had with our coral geode cache was with the pre-formed tumbled pieces. We cut the fingers and thin geodes into slices for esthetic hoops. We picked out pieces with good color and nice botryoidal surfaces and shaped them for bola ties, bracelet stones, pendants, key chains, zipper pulls, and other spur-of-the-moment ideas. We sent a huge boxful to Herb Walters of Ramona, California, who, at that time, was advertising custom tumbling. (Walters later founded Craftstones.) Each piece came back with a brilliant polish and ready to use with gold or silver for any jewelry project. We have heard that two pieces of one kind of stone are never exactly alike. Try matching any of these!

Following is a typical geode hunt in the early 1950s at the now forbidden beach at Ballast Point. The best time to be out digging was when a "blue norther" with its wild winds combined with a low tide to push the water farther back than usual. Dressed for the weather we would arrive long before low tide and would dig as we followed it out. Then with pry bars we would start feeling under the water for something hard. Sometimes we could feel the geodes with our feet. Some were covered with a gritty, bluish clay. What a thrill to pull an agatized coral from its watery grave, and what a disappointment when we were expecting a nice bottle-shaped geode to find a glass bottle instead! When the tide began to come in again we would back up to the sand and start digging a hole in what we hoped would be a lucky spot. The hole kept filling with water but we kept digging up to two or three feet and often were well rewarded. Of course there were always others there so there would be many piles of sand, muck, and debris. Since people couldn't always see what they were digging up, it paid to go back early the very next morning to see what the tides had washed for us. The small corals, particularly the fingers, were often found this way.

When we got back to our trailer in a beautiful Clearwater park, we would hose down our day's find and line the geodes up at the end of our carport. Many days in the sun, wind, and rain would make them more acceptable to take inside once we got back to our lapidary shop.

Small coral fingers or twigs are often hollow.
(Photo by June Zeitner; Zeitner Specimen)

180

Collecting agatized coral geodes in Tampa Bay in 1950 was a great adventure.

The corals weren't the only things to be hosed down. We used the hose to remove the clay, sand, and mud from our clothes before putting them in the washing machine. Our hot showers felt as good as those we took in the winter in South Dakota. A blue norther in Florida isn't exactly subtropical!

We enjoyed measuring and weighing the best corals and looking for little barnacles or clams that had accompanied the once living coral reef into a unique immortality.

FLORIDA'S AMPLE BONUSES

Tampa is not alone as a site for Florida gems. Among other sites for agatized coral are Suwannee River, Econfina River, Cross Florida Canal, Ocala, Venice, Mango, and Apollo Beach.

The dreamy Suwannee River is also home to silicified corals of several types. A typical shape of some fine examples is fan, but many exteriors look like white

The Suwanee River coral site is examined by G.B. and Ida Zeitner before he returned to dig.

grapes covered with sugary sprinkles. Much of the Suwannee River agatized coral is hollow but some is solid. The rarest and most prized is banded chalcedony in the first two layers after the outer shell and filled with vivid yellow cryptocrystalline quartz. The corals are Miocene in origin and like their Tampa relatives have occasional visitors on the exterior, bits of clams and other shells and tiny barnacles. The prized yellow gem corals are found in only one area, while other locations on the Suwannee have neutral colors, and drusy quartz linings. The Suwannee geodes are found in the riverbed when the water is low. As in Tampa, they can be located with steel rods. Some of the geodes have distinct coral markings. An unusual type has fingered hand shapes.

The silicified coral of the Econfina River is similar to that of the Tampa area except that the botryoidal lined geodes are rare, but there is a high proportion of drusy quartz crystals, sometimes in appealing pastel colors. This river is near Perry in the Florida panhandle. Also in the Withlacoochee River in far north Florida there are coral heads with cavities resembling fabulous miniature caves.

Dredging for the Cross Florida Canal brought sandy silt, and as new silt piles lined the growing canal, rockhounds discovered numerous fine coral geodes, most of them with tiny drusy quartz, appearing diamond-like in the sun. Coral heads have also been found in a lake area northeast of Zephyrhills in central Florida. They were found during the construction of a resort-type trailer park. Most of the spots where coral geodes were found in the past are now covered with concrete, grass, trees and buildings. An exception, up to a few years ago, was Perry, where a number of paper companies built along the rivers. Some of the companies would let people in to dig at times. However, I am sure that Florida collectors keep track of all new excavation and construction projects, and such events as floods or droughts.

A beautiful quartz geode with barite and marcasite inclusions.
(Photo by Jeff Smith; Jeff Smith Collection)

CHAPTER THIRTEEN

Not Always Quartz

— JUNE CULP ZEITNER —

I recently added to my geode collection with a blue barite geode from Colorado. The location is Pawnee Buttes, near Stoneham in Weld County in Northeast Colorado. It has long been noted for its tabular blue barite crystals. The barite area is extensive in a rather rugged and remote section of the Colorado prairies. Nearer to Stoneham the barite is found as lustrous tabular crystals often in clusters. There is a farm near Stoneham where collecting has been allowed. The barite geode location is farther to the west. Barite is one of the most attractive and collectable of the sulfates. Besides blue it is found in tints of yellow, yellow-orange, red, and green, and neutral grays and browns as well as colorless. Crystals are found in concretions, related to geodes. Other forms are roses, walnuts, cockscomb, and large clusters composed mostly of sand. The barite deposit has two major creeks, Pawnee and Two Mile, which flow into the South Platte River, which drains a large area of northeastern Colorado.

The geodes vary in size from 2 inches to 3½ inches. Not as hard as more common quartz geodes, they are much heavier. (Their specific gravity is 4.3 – 4.6 as compared with that of quartz, 2.65.) In places the ground sparkles with broken crystals. Parts of geodes are not difficult to find or dig out of sedimentary cliffs. Whole geodes are scarce. Light yellowish soil often has good pieces and crystals, sometimes doubly-terminated. Clear blue crystals can be found where fractures and crevices in the shale are lined with white calcite. Flat shelves on the sides of cliffs or buttes that have eroded areas around the base often have fine portions of blue geodes. Except for the National Grasslands all of this regions is private, and permission must be granted before any hunting.

Blue barite geodes from Colorado are interesting non-quartz anomalies. Tim Bachand, collector. (Photo by David Phelps)

The blue crystals are translucent to transparent and have a good luster. Some of the geodes are round, some oblong, some pear shaped, and others egg shaped. The blue may be quite pure, slightly grayed, or with the tiniest hint of green. Many geode crystals are short, thick, and tightly packed. The outer shell of the geodes is a hard, yellowish clay about ¼ inch to ⁵⁄₁₆ inch in depth. The luster is excellent. The yellow tones of the exterior seem to enhance the blue. (The crystal clusters from near Stoneham, often three or four inches in length weather out of this same yellow clay.)

Some of the geodes are almost solidly full of crystals, making even a small one surprisingly heavy. My best blue barite geode is 2¼ inches in diameter and 3¼ inches high. There is a small cavity surrounded by fourteen terminated crystals. It is shaped like a large hen's egg and about that size, but the upper part, probably about a fourth, is missing. The part I have weighs seven ounces.

The Colorado barite area does not recognize state lines, of course, so a diligent collector should take time to investigate the hills and flats southwest of Sidney, and Kimball, Nebraska, and southeast of Bushnell, also around Lodgepole Creek.

The importance of the barite deposit in northern Colorado is emphasized by descriptions in books by Sinkankas, William Ford, and the *Encyclopedia of Minerals* by Roberts, Campbell, and Rapp. Books have not mentioned these exceptional geodes, which leads me to think they are either from a later find or that they really are scarce, but they do add a new dimension to the geodes of America.

A second unusual geode turned up at our club show, this one previously unknown to me – a puzzling stranger. A Colorado collector, Tim Bachand, showed me a couple of samples of these weird geodes that he had found in his home state. He gave me part of a most intriguing one. It is white on the exterior with a thin textured shell. I believe it to be another example of geodized fossils. The fossil it resembles closely is the agatized coral of Tampa, Florida. The irregular and elaborate freeform shape is close to being a twin of some of my coral geodes. The big difference is in the mineral composition of the two examples. While the Florida coral pseudomorphs are mostly quartz, hardness 7, the Colorado geodes are the much softer calcite and gypsum with small rainbow shaped selenite

The weird shapes of some barite geodes resemble the shapes of Florida geodes.
2½ inches × 3 inches. (PHOTO BY DAVID PHELPS; ZEITNER SPECIMEN)

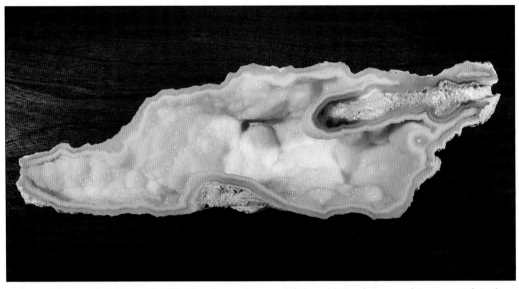

Compare this Florida geode with the shapes of some of the oddities from northeastern Colorado.
5 inches. (Photo by David Phelps; Zeitner Specimen)

crystals lining the hollow which parallels the unique freeform shape. The snowy exterior seems to be mixed with heavy clay, dull except where sparkles of selenite shine through.

Tim later gave me three more of these geodes and one of them told me where they all came from. It had a blue barite crystal inside. The location had to be Weld County, Colorado. As expected, the one with the nice blue barite crystal was heavier than the others, but the others were also too heavy for their thin shells so they probably had small barites in their elongated shapes. Tim said the area where he found these was quite remote and roadless.

SOUTH DAKOTA BARITE

The Northeastern Colorado barite district reminds me of a barite deposit near the town of Belvidere in southwestern South Dakota. A large farm field, too rough to cultivate, although it looked like this had been tried, was littered with clusters of shiny, water-clear, transparent barite crystals. Most appeared to be barite crystal nodules, although many were broken barite geodes. The short orthorhombic

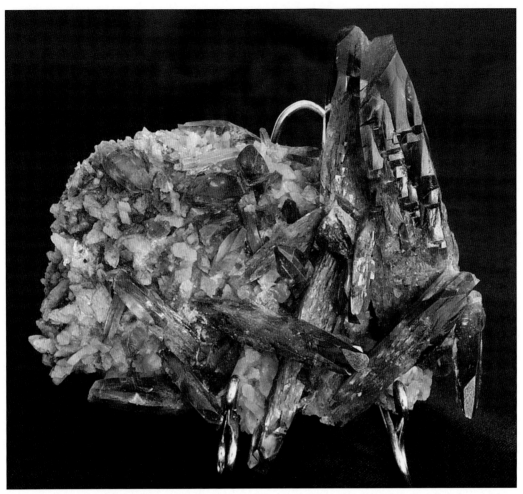

Golden barite, South Dakota's premium crystals, occur in septarian concretions, related to geodes.
5 inches × 3½ inches. (PHOTO BY DAVID PHELPS; ZEITNER SPECIMEN)

crystals had smooth gleaming faces and were often double terminated. The very thin exterior coating was always earthy yellow brown goethite. Whole geodes were scarce, but the solid crystal nodules were not. I only found two excellent geodes filled with lustrous barite crystals. I gave the best of these to my dear friend Russell MacFall many years ago, thinking I could always find another. Russ was an avid barite collector. No, I didn't ever find another fine example. The last time I tried to find the place, the roads had been changed, the landmark

*This field near Belvidere, South Dakota, was once covered
with water-clear barite geodes and concretions.*

house was not in sight, and no one I asked about the barite knew what I was talking about.

My home state, South Dakota, is noted for many forms and deposits of barite. The most famous are the elegant golden barite crystals found in Pierre Shale concretions, some of the finest barite on earth. More about them in the chapter about concretions, close relatives of geodes.

FROM CELESTITE TO IRON

Celestite is a member of the barite group. A source of strontium, it is used for its crimson red color for fireworks, flares, and tracer bullets. There are several deposits of celestite in Texas, some including geodes. The leading deposits are in

Geodized fossils, Florida geodes, and Colorado quartz-free geodes are examples of the many varieties of sedimentary geodes. 3½ inches and 5 inches. (PHOTO BY DAVID PHELPS; ZEITNER SPECIMENS)

Lampasas, Brown, Burnet, Mills, Coryell, Hamilton, and Williamson. Calcite geodes are found in some of these same counties. Travis County is a notable example.

The white and transparent crystals in geodes are glistening and beautiful but when the crystals are tinted with pastel blues or pinks they form spectacular geodes. Travis County geodes, found near Austin have square prismatic crystals with white basis and contrasting purple-blue terminations. Brown County near Blanket has geodes with white tabular celestite crystals and occasionally pale red. Other Texas deposits are in Coke, Fisher, and Nolan Counties. Calcite-celestite geodes come from the San Gabriel River in Williamson County.

An odd geode occurrence, as well as puzzling, is near Paris, Texas. It is difficult to imagine geodes like this. Maybe the residents of the Sulphur River area who reported these geodized fossils were overly excited when they described them to me. What they said was that geodes along the river lined with small calcite crystals also contained small, sometimes broken, marine invertebrate fossils.

A fascinating geode variety is found near Marfa and Alpine in Presidio and Brewster Counties. We found ours at a ranch in Presidio County in the agate rich land that stretches south into Mexico and east to Big Bend National Park. Although it is an igneous agate, I thought it deserves describing here among the geode oddities of Texas. Texans, and sometimes others, think of Texas as having everything oversize, but this example is decidedly opposite. The tiny drusy quartz geodes, from ⅛″ to ⅜″ in diameter are entombed in translucent agate. The agate is gray, rusty-red, brown, orange, and yellow. No matter how thin or thick these agates are slabbed, several of these little geodes will be cut open, because they are so crowded in their agate home that each one touches four or five or more of its neighbors. Each crystal pocket is bordered by a chalcedony rim of one of the agate colors and outlined in yellow. I think one of the little geode halves in agate centering a cabochon would make an interesting pendant or pin, an exotic use for a geode.

Oklahoma is another geode-rich state. My late friend Marie Kennedy, who had an exceptional collection of self-collected Oklahoma and Kansas minerals, had fine geodes from both states.

A west Texas conglomerate agate is made up of tiny geodes, mostly under ¼ inch.
Slab 4½ inches. (Photo by David Phelps; Zeitner Specimen)

Hollow brown iron mineral nodules with loose sandy crystals inside, found in Wyoming, Texas and other states, have often been called iron rattle-rocks.
(PHOTO BY DONNETTE WAGNER; BRAD L. CROSS COLLECTION)

From Comanche County came the geodes Marie called, "bunch of grapes." Inclusions are calcite and barite. These geodes are brightly fluorescent. Calcite and barite geodes also come from Cotton County and McClaine County. Keokuk-type quartz geodes are found in Delaware County. Marie kept a few of the large and seldom mentioned ironstone geodes from Rogers County. (We once found small ironstone geodes in Wyoming near the Black Hills.) Perhaps the most unusual of Oklahoma geodes are those of Washington County. Some are solidly full of scenic calcite and have been cut and polished by local lapidaries.

Kansas has geodes along the Walnut River in Butler County. Good quartz geodes come from Chase County. Other geodes are found in the limestone beds of the following counties: Logan, Cherokee, Wallace, Trego, Riley, Marshall, and

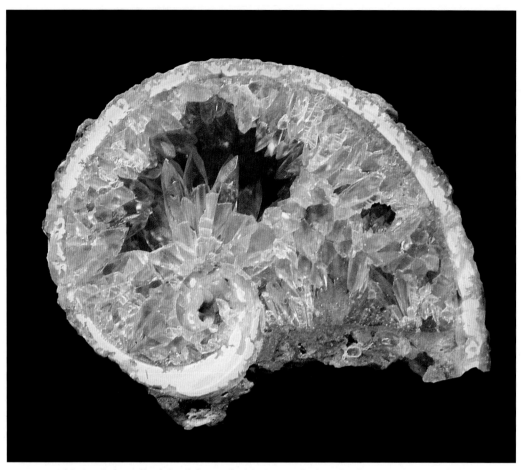

This partial geodized fossil from Florida is a cephalopod. The crystals are calcite.
(PHOTO COURTESY OF FLORIDA GEOLOGICAL SURVEY)

Cowley. The Cowley location, also on the Walnut River, is near a town appropriately named Rock. A third Walnut River deposit is near Douglas in neighboring Butler County. All the geodes are chalcedony, quartz, and calcite, with few inclusions. Kansas is also noted for many concretions, related to geodes. Some are really giants as told in the concretion chapter.

Virginia has geodized fossils like those of Kentucky, Tennessee and other southeastern states. Near Virginia Beach, hardened fossil clamshells reveal shining ivory colored calcite crystals. Another geodized fossil site is near Saltville in Smyth County.

South Dakota barite geodes are quite a contrast to Colorado's. 2½ inches.
Black Hills Museum of Natural History. (PHOTO BY LAYNE KENNEDY)

West Virginia's Mineral County has small chalcedony geodes with lucid quartz crystal linings in an inactive quarry near Berkely Springs. In North Carolina small amethyst geodes are found in Rutherford County in a scenic area not far from the Blue Ridge Parkway.

A visitor to an iron mine in Alabama once found a large and heavy ironstone geode. Mystified by the unusual rock he asked about it in the mine office. The busy worker replied quickly. "Oh, what a nice black diamond!" The elated novice rockhound rushed off to show his incredible find and to write an ad for publication offering to sell it for one million dollars. He also booked several shows where he could put on a special display. Sometime later someone had the nerve to tell him that "black diamond" was only an inappropriate name for hematite, an iron ore. I can only guess at his sorrow and chagrin, but I do know he retracted his ad and apologized for the confusing stories that appeared. I saw the "geode" in a brightly lighted case but not close enough to tell whether it was really a geode, or an anomaly, or part of a variation of kidney ore, the product of weathering, or even a plating over another iron mineral. It was jet black and had a brilliant metallic luster.

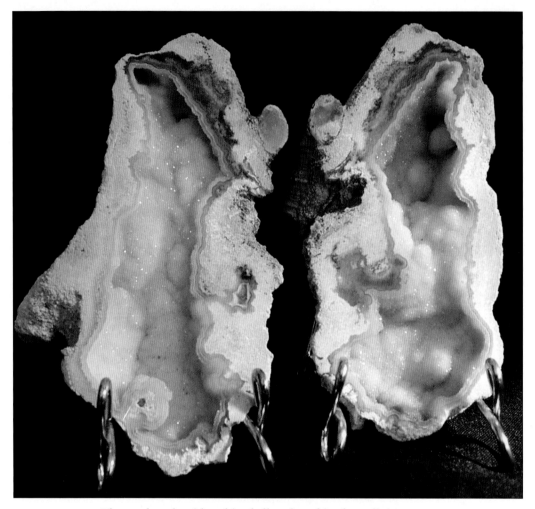

The coral geode with a thin shell and a white drusy lining was a
Suwanee River enhydro accidentally cut by a lapidary.
3 inches × 4¾ inches.

(PHOTO BY DAVID PHELPS; ZEITNER SPECIMEN)

CHAPTER FOURTEEN

Strange, Stranger, Strangest

— JUNE CULP ZEITNER —

Water filled geodes, enhydros, are the most common of the geodes containing liquids, semi-liquids or gas. Water is found in both igneous and sedimentary geodes. In both cases the water is enclosed in quartz crystal lined cavities. In sedimentary geodes, the amount of water varies from about a teaspoon full to a quart or more. When the geode shell is thick or milky in color the water may be a complete surprise when the geode is broken or cut. In some specimens the water can be heard. It can be observed in the sedimentary geodes only if the shell is thin translucent chalcedony or agate, for instance the Tampa Bay Florida agatized coral geodes.

An air space is usually present with the entrapped water. This space may signify the evaporation of some of the water. In thin-shelled translucent geodes where a partially empty cavity can be seen there may be other gases. In Swaziland, Africa, some have been found with nitrogen, oxygen, and carbon dioxide. In Victoria, Australia, sulfate and chlorite in solution have been found. Good examples of enhydros in the United States come from Lincoln County, Oregon, San Mateo County, California, Clark County, Missouri, and Hillsboro County, Florida.

Clifford Frondel wrote in his 1962 edition *of Dana's System of Mineralogy, Volume III* of geodes containing several hundred milliliters of water. The enhydro phenomenon was noted by Agricola who described them back in the early sixteenth century as being, "round, smooth, and full of liquid."

Hollow quartz casts that contained liquid have come from two counties in North Carolina where empty broken drusy quartz geodes were found. Broken by

White or light-colored geodes with translucent shells are often enhydros.
(PHOTO BY MILLIE HEYM; HEYM COLLECTION)

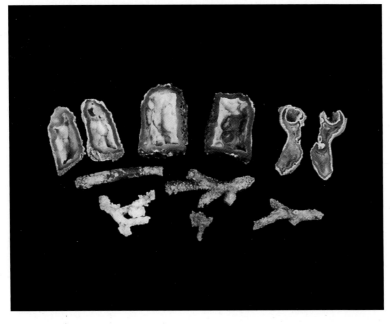

Even small coral fingers may hold water.
(PHOTO BY STAN KAPPIRIS)

nature, they probably held lots of water since some of these super geodes measure two to three feet across.

Water is almost always present in quartz, though most of it is invisible or microscopic and sometimes barely visible to a determined researcher. Cryptocrystalline quartz is quite porous. If crystalline quartz is snowy white it probably contains more water than most other varieties of quartz. The largest amounts of water are seen in enhydro geodes, where transparent quartz crystal linings enclose the water. The water may have been entrapped in initial phases of growth.

After being enclosed in rock it may be almost pure or saturated with potassium, sodium, carbon dioxide, calcium, or chlorine. The water enclosed in the numerous Florida enhydros I studied reeked of sulfur. The dissolved material in geodes ranges from a very small percent by weight to as much as thirty percent. Several magazines have printed pictures of people drinking from newly opened Keokuk geodes. Although the imbibers claimed the waters to be wonderfully refreshing, some of the countenances looked less than happy. Most enhydros dry out soon after they have been removed from their native rock formation. When we were first hunting Tampa geodes we were told that if the water could still be seen and/or heard after five years, it was probably ancient water. Some of our enhydros were still water filled after fifteen or twenty years. Most were apparently empty.

Most of the best enhydros for display currently come from Brazil. Some of these have translucent chalcedony walls. In others the walls are thick enough that the enhydro may be carefully polished in the appropriate spot. Polyhedroids, also from Brazil, are often enhydros, but since they are more interesting when cut and polished, the water is lost in the process.

Nature's water bottles, enhydros, are almost hypnotic. As you tip translucent ones back and forth in the light you can lose track of your surroundings wondering how many millennia ago it was entrapped, and there it is, crystal clear and – according to those who have tried it – "potable."

Most ancient water has been found to have less mineral content, salt included, than present day seawater. Not all geodes holding water are true enhydros. In some cases the geode shell may have small holes or fractures so water could enter it. If a geode is suspected of being an enhydro, it should be kept in a dry place for several years, and if after five years or so it still has the same amount of water it is considered an enhydro.

Although many enhydros have been found at Ballast Point, Florida,
Warsaw Formation areas have also yielded good examples.

Lapidaries sometimes polish enhydros, especially those from Brazil, when the chalcedony is translucent to transparent and thick enough that careful techniques will not break it. Obviously it cannot be exposed to much heat. Good natural enhydros are treasured by rockhounds, but recently I have heard of some being faked by clever sellers.

We had a thin-shelled translucent water-filled enhydro from Tampa, Florida – one of those coke bottle-shaped ones. We wired each end with gold wire and suspended it in a small glass display case so we could mechanize it to be constantly in motion when the current was on. Museum visitors seeing something like this for the first time were transfixed. It was one of our most popular exhibits. A good-natured banter usually ensued when I referred to it as "fossil water." If the water was incorporated in the geode when the calcium carbonate became quartz, I referred to one of several meanings of fossil – ancient, prehistoric, or primordial.

Some people keep their enhydros in containers of water, preventing them from ever finding out if they have true enhydros or not. If an enhydro appears to be genuine, it need not be put in water, but it should be kept from freezing and any kind of rough treatment.

BITUMEN

Among the most unusual geodes anywhere are those from Niota and Nauvoo, Illinois. These strange geodes are filled with bitumen, a viscous hydrocarbon variously called petroleum, asphalt, or oil. The prime location is along Tyson Creek, although a few have been found farther south toward Hamilton.

We had read about these geodes years ago in the *Mineralogist Magazine*, so in the late 1950s, we decided to try our luck in this historic area along the Mississippi. Friends from Iowa had drawn a map for us, so in a short time we could spot geodes on both sides of the

Dean Thompson shows a Tampa drusy quartz coral geode cut with his trim saw. The geode turned out to be an enhydro. 5 inches.

pretty stream and its banks. Most were rather small on average, compared to geodes of Iowa and Missouri. The larger ones we spotted were only about seven inches in diameter. Many of the geodes were stained gray or black, but the stains were dry and hard.

Albert, ever curious, opened one on the spot. Out dripped some gooey, smelly, thick crude oil. As the black liquid slowly slipped into the measuring cup we could begin to see the terminations of quartz crystals which lined the geode before it met with this weird Halloween trick. There was only a little over half a cup of petroleum. We decided not to go into the refining business. Almost as much as a pint of oil was reported by residents who had found larger geodes in the past.

The Tyson Creek geodes weather out of loosely consolidated limestone. We had read about them since the thirties and when we went there in the late 1950s, they were still plentiful and could be seen in-situ. These curiosities have been available for many years. Dr. Carroll of the Illinois Geological Survey wrote that the oil had to be present in the limestone cavities before the geodes were formed. I noticed that most of the Niota geodes had thicker walls than the average Keokuk region geode.

How did the oil get into the geodes? The question puzzled geologists and collectors back then, and it still does. Oil deposits were formed in the Pennsylvanian period from decaying biota. The Pennsylvanian span was 25 million years and followed the geodiferous Mississippian period of 30 million years duration. A theory that the oil seeped down through these thick formations into the geodes is debatable, partly because the nearest known Illinois oil deposit is many miles away. The quartz crystal linings of the geodes must have been formed before the oil entered. How could the oil get into geodes with no fractures or holes? Why isn't it possible that since oil is a product of decay, some of the geodized fossil type of geode incorporated Mississippian marine remains as the geodes formed. Would not the soft marine life become oil during those millions of years? In that case the black stains on the Tyson Creek geodes could have come from tiny pores in the inside of the geode rather than dripping and drifting from oil deposits far away.

Geodes housing some type of oil (bitumen, tar, or petroleum) are not common. However, there are known sites for these occurrences on several continents – Asia, Australia and North America. Years ago, a friend who worked on the Panama Canal while it was maintained by the United States told me of finding oil-filled geodes in Panama. I have not been able to get verification on his information. I do know he had some Keokuk-style geodes and Florida coral geodes in his collection.

Not far north of San Francisco is a small city named Bolinas with a beautiful Pacific beach where oil filled geodes called "Bolinos" have been found. The

Tyson Creek, Illinois, has been known for its oil geodes for many years. This one is about 3 inches square.
(PHOTO BY DAVID PHELPS; ZEITNER COLLECTION)

chalcedony "oil cans" are sometimes almost glassy clear. Most are translucent. The black tarry inclusions are easily seen.

Eager collectors, including some of the California gem and mineral clubs, dig the geodes from the beach sands.

The location is Agate Beach, a scenic area of the California coast near the Bolina Lagoon. The "oil agates," as they are called, are easier to find after a high tide or a big storm. They are quite round and smooth with black blob inclusions seen easily in the transparent to translucent chalcedony. Agate Beach, the main deposit, is a park at the end of Elm Road in Bolinas. A parking area is located by the Pacific. Jasper is also found on Agate Beach. Similar geodes are occasionally found farther north.

Another inclusion in geodes is responsible for the rotten egg smell associated with specific specimens in certain areas. The unpleasant odor is from hydrogen sulfide gas. Deteriorating pyrite or some other sulfide is thought to be the culprit. Odorless gasses, oxygen for example, are also included in geodes.

POISON

As interesting as inclusions of water, oil, and gas are in geodes there is another I find even more extraordinary. It is liquid. It is used in industry. It is a poison. It is mercury.

Stranger geodes than those of Napa County, California, are difficult to imagine. These geodes include the native element mercury. It is used in thermometers, barometers, arc lamps, and alloys such as processing gold ore. The common name, quicksilver, is descriptive. The brilliant silvery heavy liquid is also used in dentistry.

Native mercury is relatively rare. Besides Napa County in northern California, there are deposits in nearby counties Sonoma, Lake, and Santa Clara. Mercury has no resemblance to any other native element or mineral. The mercury-bearing Napa geodes were found in the Pioneer Mine and were first mentioned in *Dana's Manual of Mineralogy* in 1895. The unique geodes weathered out of stratified limestone in the old mine.

Geodes with some kind of inclusion are more interesting and collectable than the much more common chalcedony/quartz variety. The three types with liquid inclusions – water, oil, and mercury – are puzzling, fascinating, debatable, and sometimes beautiful.

These are by no means all the common and uncommon geodes we have covered in this book. In fact, in our research we have found that the extended geode family is much more extensive and that geodes are a lot more mysterious and puzzling than we first thought. There are numerous theories and debates among scientists that are far from being resolved. An example is the geodized fossils. It was not long ago that some geologists said the very idea was preposterous and that these strange geodes merely *resembled* marine invertebrates. Geometric geodes like the polyhedroids were never even imagined. I certainly never dreamed that some of the most esthetic geodes were only a short distance from my home. I found many papers and books that referred to vugs and geodes as being one and the same! And as I am writing this I keep looking at a geode beside me, different from any I have seen before. I have collected sedimentary geodes in fifteen states and bought a few from other states, so I thought this book would be easy. Far from it! But I'm glad I did it. I used to be just mildly interested in geodes, mostly the gorgeous ones like the amethyst geodes from Brazil. Now I am a full-fledged geode enthusiast, looking at every one I see as a chance to learn something new.

Part Three

Thundereggs and More

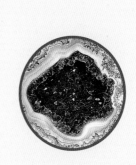

CHAPTER FIFTEEN

Dozens of Eggs

— BRAD L. CROSS —
— JUNE CULP ZEITNER —

Unrelated to the weather and not affiliated with chickens, thundereggs are an interesting problem. There is some confusion about geodes and thundereggs – they are not the same things. A thunderegg is a special kind of spherical to egg-shaped nodular mass of rhyolitic material that occurs in perlite or decomposed perlite beds and in the glassy portions of welded tuffs. Perlite is a hydrated, silica-rich volcanic glass. A welded tuff is a rock that forms when volcanic ash remains hot enough for the glass shards to almost melt togther. The nodules are approximately spherical in shape and have an exterior shell of rhyolite that is more silicified than the host rock. The term thunderegg refers to this special type of nodular formation including its interior. Thundereggs may be hollow (geodal thundereggs) or filled with agate, jasper, opal, and other minerals.

Where did they get the weird name? According to reputed legends of the Warm Spring Indians of central Oregon who dwell within the shadows of the volcanic peaks now named Mt. Hood and Mt. Jefferson, the angry mountain gods waged war on one another by hurling eggs gathered from the nests of thunderbirds. When the gods hurled these geologic missiles or "thundereggs" across the skies from one mountaintop to the other, errant shots sometimes fell on tribal land, thus accounting for their presence there. When opened, they reveal quite interesting and intriguing interiors.

Today, thundereggs are very popular as collectors' items, and tens of thousands are sold every year to tourists, particularly in Oregon. In fact, the Oregon state rock is the thunderegg, even though technically it is not a rock. Most thundereggs are about the size of a baseball, but they can be as small as a pea or as

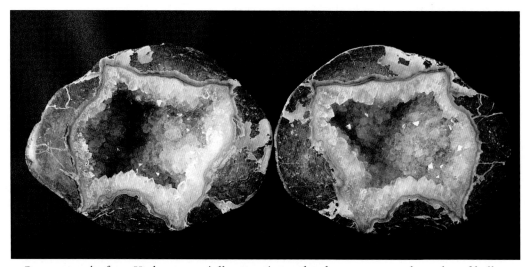

Dugway geodes from Utah are especially attractive as they have an unusual number of hollow centers. Because of their irregular rhyolite frames, they are classed as thundereggs. 5-inch specimen. (Photo by David Phelps; David Phelps Collection)

The internal core of a thunderegg. Each side is referred to as a "button." In the center of each button, a "dimple" (indented area) or a "pimple" (a raised area) will be found. (Photo by Donnette Wagner; Brad L. Cross Collection)

large as several tons. Some criteria for recognizing thundereggs include: (1) a blocky to geometric internal structure; (2) raised "ridges" around the outer surface; (3) a generally warty exterior; (4) brecciated non-chalcedonic, rhyolitic inclusions; and, (5) impressions of radially arranged quartz and feldspar crystals.

The central portion of a thunderegg is referred to as the "core," and can be small or large. The core is usually completely filled with agate or opal and can often be lined with small, crystallized minerals. Each face of the core is depressed inward to the center. A fibrous structure cast on the agate cores radiates from the center of each face towards the edges. Often the top and bottom faces of cores have, on one face, a depressed dimple at its center, while the opposite face has a raised "button" embossed at the center of its face. The button and depression on the top and bottom, not always respectively, is suggestive that these two opposite faces were once common together and had been forced or drawn apart. Thundereggs are readily identifiable by their internal angular shapes or patterns, while geodes usually have a continuous round interior. Thundereggs are not always completely filled and may be hollow. These hollow structures are frequently sold as "geodes."

Thundereggs by other names were documented early, in Europe at least well back into the eighteenth century. They were described as spherulites in 1891 by the American petrologist, Whitman Cross. They remained largely unknown until the mid-1930s when articles in *The Mineralogist* drew attention to the Oregon thundereggs. These deposits are most commonly associated with welded ash flow tuffs in the John Day and Clarno formations of Miocene age, in the Antelope – Ashwood areas of central Oregon. With their wide distribution, discoveries were soon made elsewhere in the western United States. Other well-known deposits are documented at Dugway, Utah, and near Rockhound State Park just outside of Deming, New Mexico.

Thundereggs usually have a rather irregular, bumpy or wart-like surface, sometimes with intersecting ridges that encircle the nodule. While the ridges on the outside of the thunderegg may appear randomly distributed, they give an indication of the internal structure. These ridges represent the point at which the internal agate filling reaches the exterior surface of the nodule. These features may be more prevalent in one thunderegg than another. Four different types of thunderegg interiors have been identified which probably represent different degrees of temperature and pressure in the original formation as well as differences in original magma composition. One produces a single lense cavity, another

Many thundereggs contain outer "ridges," representing the outer perimeters of an internal core of agate.
(PHOTO BY DONNETTE WAGNER; BRAD L. CROSS COLLECTION)

produces a triangular cavity, the third a box-shaped cavity, and the fourth the classic star-shaped cavity. Practically every thunderegg-producing site yields it's own uniquely structured and patterned features.

In the field, a thin zone of porous material surrounds the thundereggs. It is presumably derived from a low-temperature alteration of the host rhyolite that produces clay and/or zeolite minerals. The thunderegg shell is composed of silicified, spherulitic, rhyolitic material that is considerably denser and more weather-resistant than the host rock.

Various theories have been offered to account for the manner in which thundereggs grow and develop. The geometric form of a filled cavity that will display a five-pointed star in cross section presents an interesting problem. Volcanic activity sometimes produced a rhyolitic lava flow and sometimes a glassy ash that fell in a hot plastic condition, permitting its welding into a nearly homogeneous material resembling a lava. In both cases, local centers of crystallization were set up, around which spherulitic masses of inter-grown cristobalite and feldspar were formed. The formation of these minerals released gases originally in solution in the glass. As crystallization proceeded, more and more gas collected. This gradual collection of gases exerted a pressure, which combined with the cooling shrinkage of the enclosing material, forced the walls of the cavity outward. Rhyolitic lavas and portions of ash falls destined to become welded tuffs are highly viscous, so expansion by rupture, or tearing, required much less energy and therefore rupture occurred along symmetrically arranged planes.

Ross (1941) demonstrated that the more perfect of these cavities have the geometrical symmetry of a modified pyritohedron bounded by twelve inward-projecting, five-sided pyramids, formed by the shear along thirty triangular planes. A figure such as this will give a five-pointed cross section when cut in a number of directions, the most perfect star being given when the cut is made obliquely about one-third above or below the equator of the spherical nodule. In nature, there will be various modifications and distortions due to lack of

McDermitt, Nevada, thundereggs include some with reddish rhyolite and multiple pointed stars. 4 inches. (PHOTO BY RUSSELL KEMP)

homogeneity of the lava or ash that will modify or destroy the ideal symmetry. The irregular ridges or ribs on the outer surface of the thunderegg represent the boundaries between the twelve pseudopyritohedral faces and outline twelve approximately equal areas.

Some thunderegg names combine the word "thunderegg" with a well-known geographic locality. For example, collectors around the world recognize the term Priday Ranch thunderegg. Many other terms, however, are only locally understood (or misunderstood). Still other terms are the creations of dealers and hobbyists, who coined these names in hopes of selling their so-called thundereggs. These commercial names are rarely defined, rarely are given a location, or have an adequate description. When a thunderegg is hollow and thus a geode, the geode term usually takes precedence. Thus we have Dugway geodes and not Dugway thundereggs. Appropriate or not, these terms are with us.

UNITED STATES

Oregon

There really are dozens of eggs in Oregon. Thundereggs, that is. There are also dozens of locations for these geode relatives, probably more than any other state. The T-eggs from various sites have different features and are named for their locations, Ochoco, Succor Creek, and Madras for instance.

Larry Krebs, miner, collector and dealer, claims Oregon has more than fifty known deposits, and maybe many still undiscovered. According to his count in the late twentieth century there were thirty beds near Madras alone, plus twenty in the Ochocos and others near Buchanon, Nyssa, Steens Mountain, Succor Creek, Crowley, Lakeview, Maupin, Pony Creek, Ironside, Dry Creek, and Prineville. The beds are found in most parts of the state east of the Cascade Mountains. A few beds are found on the west side of the Cascades.

To me, the Priday Ranch thundereggs are the most beautiful of any; not every one of course, but the best have few peers, and none rate higher. Albert cut many of these; and occasionally we would be awed by the beauty of the plume agate bouquets of lifelike flowers, pink, lavender, rose, cream, yellow and

Ashwood, Oregon, is in a thunдеregg-rich locality.
Dark rhyolite and onyx banded agate are common varieties here.

Butte Creek, Oregon, is famous for large thundereggs with warm brown patterns. 14 inches.

213

Strata as exposed by strip mining at the Priday Blue Bed at Richardson's Recreational Ranch, Madras, Oregon. The thundereggs are found below a dense layer of rhyolite.

white against a true blue chalcedony ground. Sometimes the plumes looked like a flowering crab tree when the blossoms all come out before the leaves. I have seen choice 30 x 40mm cabs of Priday eggs sell for $400. (I'm glad agates are no longer miscalled *semi*-precious!)

Richardson's Rock Ranch operates the famous Priday thunderegg beds and offers excellent digging opportunities. The ranch is located eleven miles north of Madras on U.S. Hwy. 97.

Other interesting Oregon T-eggs are those with multiple agate pockets surrounded by multiple jasper-like stars. Then there are picture agates in the center of some eggs. One in the Zeitner collection looks exactly like a miniature Scotch Terrier. (A portrait of the world's oldest rock-hound?) Another resembles a ship with all sails unfurled. A favorite resembles a scene in a tropical coral reef. Thundereggs are exciting and fun!

Blue common opal occurs in thundereggs in the Desert Dog Mine near Bend. It is fine cabochon material. The claims are in the Stockade Mountains. The owners report an occasional thunderegg with precious opal.

A Prineville thunderegg features a picture of a ship at sea. 4¾-inches. (JUNE ZEITNER SPECIMEN)

In many of the West Coast T-eggs the rhyolitic outer portion, distinguished by its broken star-like frame, has a top quality translucent blue chalcedony halo-like layer between the agate and the frame.

The opaque rind of the cut thunderegg will polish also, so many thunderegg cabochons incorporate portions of the darker opaque material to accent the pastel colors of the agate portion.

To me the most glamorous of thundereggs are those from Opal Butte, south of Heppner in Morrow County. These exotic thundereggs contain precious opal. The opal site was first reported by Kunz in 1893. The opal-filled eggs are found in light colored clay. The specimens with the highest percentage of gem opal are usually between fourteen and eighteen inches in diameter. Only about one in twenty of the Opal Butte T-eggs has precious opal, but what marvelous and valuable opal this is. Some of it is almost transparent. The captivating play of color is *contra luz* (against the light). The basic color is often blue and the flashes of changeable colors are rose, cerise, green, yellow and violet. Other eggs contain hydrophane – opal which absorbs water and becomes transparent when wet and sometimes shows a play of color. Common opal is found in other specimens. The

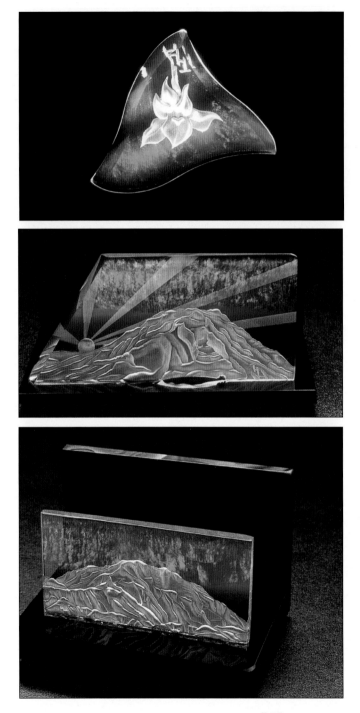

Artist Thomas Ames uses contra luz precious opal from Oregon's Opal Butte thundereggs for his award-winning carvings.
(PHOTOS BY MIKE BUSH)

best of the contra luz opal from the Opal Butte Mine has sold for $50.00 per carat. Lapidary artist Thomas Harth Ames has won numerous awards for his creative carvings of this mangnificent opal.

Idaho

The mountains of Idaho have several beds of thundereggs. Eggs of all sizes filled with fine agate and jasper and even common opal have been found south of Twin Falls, along the Nevada border. Succor Creek Canyon, close to Oregon, has a different T-egg offering. Instead of banded or patterned agate, the eggs may be filled with blue agate, common opal or rarely precious opal. The famous common and precious opal of Spencer occurs in thundereggs.

Bruneau jasper, one of the best known of Northwest jaspers, occurs as thundereggs some fifty road miles south of Bruneau in Bruneau River Canyon in Owyhee County. It is noted for its attractive colors, which although monochromatic and neutral, look as if they had been patiently mixed by an exacting artist. Cream, buff, beige, tan, rusty red, and several shades of brown are in perfectly drawn circles, arcs, ovals, ellipsoids, or egg shapes. Graceful ovals are light cream

Idaho thundereggs often have good blue colors. Double and triple centers are not uncommon. This one has either four or five. 6½ inches.
(PHOTO BY DAVID PHELPS; JUNE ZEITNER SPECIMEN)

in the center and gradually deepen in color until a distinct red-brown outline appears. Bruneau jasper has been cut for many years into jewelry, novelty and display items, and specimens.

Another popular lapidary material, Willow Creek Jasper, occurs as thundereggs and is found approximately fifteen miles northwest of Boise. Some of the thundereggs are as large as ten feet in diameter. Willow Creek Jasper is known for its subtle pastel colors, streamer patterns, and occasional egg patterns.

A principal Idaho thunderegg area is along Poison Creek in Owyhee County in the southwest corner of the state. Many "eggs" taken from here are solid, but those with cavities have quartz or calcite crystals or varied zeolites. These eggs run to "extra large" size.

California

Thundereggs are found about forty miles south of Blyth, Imperial County. The location in the Little Mule Mountains yielded a record specimen 42 inches in diameter and over 2,000 pounds in weight. The Chocolate Mountains of Imperial County also have thunderegg deposits. The Hauser beds near Wiley Well have long been popular. The eggs are colorful and have a variety of agate patterns including banding, mossy inclusions, and straight-banded onyx.

The Chuckawalla Mountains of Riverside County have a T-egg deposit south of Desert Center with vivid examples, many of them with crystal linings. The Mojave Desert also had productive sites northeast and northwest of Barstow. Thundereggs weathering out of disintegrating rhyolite were discovered in 1930, however, the area was soon covered by homes, industry, and roads.

Among the most beautiful of California's thundereggs are those from Lead Pipe Springs, now a closed locality. These Mojave Desert beauties hide high-quality blue chalcedony, one of the most sought after chalcedony gems. The rhyolite exterior is a rich red color, a spectacular contrast to the vivid blue. The T-eggs are now enclosed in a Military Reservation.

Utah

Geodal thundereggs and agates are plentiful near Dugway. The eggs have extreme variations in size. They have irregular shapes and rhyolitic appearing shells or rinds in neutral tones that appear to be more silicified than typical T-eggs. The

Fortification agates are among the most popular of Dugway geodes. 4½ inches.
(PHOTO BY DAVID PHELPS; RUDNEY COLLECTION)

hollows are not rounded as in ordinary geodes, but are unusual artistic shapes following the unique outlines of the thunderegg-like shells. Centers are often beautiful blue-white drusy quartz covering botryoidal surfaces. The skin of these "geodes" as they are usually called is light in color with a rough texture. Easily cut and polished they often reveal picturesque stalactites, blue chalcedony fortification agate frames, and other wonderful details. Some of the nodules are just rhyolite balls having no central cavity containing chalcedony or other filling.

There are several areas to dig near the Dugway Proving Grounds. The area is remote desert with roads to the beds fairly good except in case of heavy rain. The digging is difficult, especially during the 100° heat of summer. Roads to the beds are from Vernon, Jericho, and Delta. All the roads are about fifty miles from a main highway to the beds. The beds are not far from the famous Topaz Mountain.

Large areas of these western states are arid or semi-arid. What looks on a map to be a major river may be bone dry. Trails to collecting areas are often poor to non-existent. Those who have not yet let their hobby lure them into desert collecting should be aware that cars built for Interstate Highways are not suitable. Gas and water may be many miles away from good places to prospect. Outdoor activity in the desert heat of summer is dangerous. The good news is that warm sunny days in the winter make even a dry run pleasant.

Utah is also known for the sensational septarian concretions of Alton and Orderville. Septarian nodules are sometimes confused with geodes, but are only similar in some ways.

219

Dugway geodes. (Photos by Donnette Wagner; Brad L. Cross Collection)

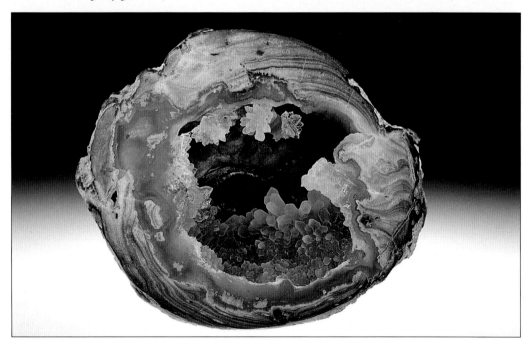

Nevada

A spot famous for its blue opal filled thundereggs is Coyote Springs in Humboldt County. This has been a good area for many years, and is still productive occasionally if the collector is willing to work a little longer and spend more time exploring and digging.

Nye County farther south in the state has an occurrence of moss agate filled thundereggs. These are unique in that the so-called "moss" can be green, red, or orange in addition to the usual brown and charcoal tones. In the Trinity Range near Lovelock in southern Pershing County several varieties of thundereggs are found. Most well known at present are the thundereggs from the Black Rock Desert area in Humboldt County.

Collectors should be aware that a large portion of Nevada, as well as sections of many other western states, is controlled by the U.S. Government. Many places formerly open to collectors are now closed. It is up to an individual to determine if collecting is legal in the spot he or she wishes to visit.

Arizona

While exploring west and south of Flagstaff, we found a few agates and some jasperized wood – then nothing of interest. We were about to give up and return to our International Travelall, when we picked up a broken thineregg. While it certainly was not a winner, it gave us new hope. We soon found several more. They were rather small and quite smooth, and reddish on the exterior. Suddenly we saw a stake, and beside it a neat pile of T-eggs similar to the ones we had been finding. We thought maybe the old badly deteriorated stake had once held a tobacco can with a claim inside, so we added our few to the pile, and started back to the agates. (It was fun to find the eggs anyway.)

Arizona is a mountainous state, mostly sparsely populated except for a few major metropolitan areas. There are most likely several thuneregg deposits in the state. We heard of one near Globe. Arizona is a state rich in gems and minerals and most collectors are interested in the vivid "petrified" wood, the finest turquoise and gem chrysocolla, the desert roses and the Apache tears, or the fine mineral specimens, so T-eggs are not close to the top of Arizona lists.

In mountainous desert country near Aguila there are thundereggs to be found in dry washes. Occasionally the dry washes become flooding rivers. Most of the T-eggs are rather small and have crystal-lined centers. A few small T-eggs were found near Wickenburg when we were spending the winters there years ago. An old location is north of Payson along the Mogollon Rim. They are dark, small and have agate interiors.

We once found a few thundereggs near Miami and Superior when we were picnicking under some old sycamore trees. When we went back to explore the area a few years later, the road had been changed and we couldn't find the same spot.

New Mexico

New Mexico is a great state for rockhounds. Noted for Luna agate, Apache agate, carnelian, wonderstone, turquoise and many wonderful thundereggs and geodes, the state is the only one of the fifty to have a "Rockhound State Park." Rockhound Park has some large thundereggs with green agate centers. The vast park, set aside in a rugged and remote mineral rich area, has many collectable items, and facilities for rockhounds to camp and enjoy a rich area where they don't have to worry about who owns and controls the land. In addition to the T-eggs in the park, more thundereggs are found all the way south to Mexico and beyond.

Large deposits of colorful thundereggs are found thirty-five miles southwest of Deming at the Baker Ranch. The exteriors are brown or red-brown, but the irregular star-shaped or butterfly-shaped outer third or more of dark unpatterned jasper-like rhyolite surrounds unexpected beauty. The agate center is usually vivid fortification bands of intense clash colors such as red, orange, black, violet, pink, and yellow. The colors many times insensibly blend into each other and are among the best found anywhere.

The eggs average about two to six inches in diameter. One-foot specimens are exceptional. Doubles are not uncommon. Some eggs have larger hollows and crystal linings, occasionally amethyst. If the eggs are filled with agate, the banding may be straight, or onyx agate. The New Mexico eggs are usually sawed and polished for display.

Perhaps one of the most experienced collectors in this region is Robert "Paul" Colburn, locally known as "the geode kid." Paul and his partner Christopher Blackwell have collected thundereggs in this region for over thirty years. Their extensive thunderegg collection can be viewed at the Luna

Baker Ranch nodule. Luna County, New Mexico.
(PHOTO BY DONNETTE WAGNER; BRAD L. CROSS COLLECTION)

Mimbres Museum in Deming. It is not to be missed! Collectors should also visit the Basin Range Volcanics Geolapidary Museum located just outside of Rockhound State Park.

Colorado

Some fine large thundereggs weighing up to over three hundred pounds occurred northwest of Del Norte. There still are a few to be seen weathering out of volcanic rock. The centers are translucent to transparent chalcedony of common opal, and occasionally moss or plume agate.

In the rugged Wolf Creek Pass area south of Pagosa Springs, thundereggs lined with quartz crystals have been collected. Some of the crystals are light

Some large orange thundereggs have brecciated patterns. 16 inches.

amethyst. Banded agate is another inclusion.

There are T-eggs in the rhyolite peaks near Saguache. Agate, chalcedony and jasper help make this an interesting area. The thundereggs have a red-brown shell and may have linings of white botryoidal chalcedony.

There are probably more Colorado mountains with deposits of thundereggs. With many rocky peaks over 14,000 feet, Colorado is the top of the forty-eight contiguous states. Digging is difficult, but summers in the Rockies are delightful.

Alaska

Large areas of Alaska are true wilderness. There are lots of mountains and lots of rocks, but few roads. There are undoubtedly many millions of collectable rocks and minerals in areas which seem to beg exploration, but the only way to get to such sites is by bush plane. A few years ago a large deposit of very attractive green marble was found in the Talkeetna Mountains, but mining it must have presented too many problems, because the material was never advertised.

There is one well-known and much visited thunderegg deposit on Glass Creek. The claim currently belongs to Mary Cary of the Chugach Gem and Mineral Society. Glass Creek is in the Sheep Mountain area of the Talkeetnas

Glenn Highway, Alaska 1, leads south-southeast of Anchorage to Tetlin Junction and passes many places worth exploration. The Glenn Highway is a route which goes close to several productive thunderegg sites. Muddy Creek north of this highway and west of Watchtower Inn has yielded fine thundereggs for years. Caribou Creek, near the bridge by Mile 107 on the Glenn Highway, can also be followed for thundereggs.

North of Chickaloon along the Chickaloon River Trail toward Puddingstone Hill (I love these names) several varieties of T-eggs are found, as well as agate, jasper, silicified wood and other collectables. A rugged roadless area of Boulder Creek south of Chitna Pass has thundereggs and geodes. This is in splendid mountain country north of Mile 70 on the Glenn Highway. South of the highway where it crosses the Matanuska River, thundereggs occur on Kings Mountain. More are found on Mazurka Creek in the Talkeetna Mountains. Chickaloon is just east of the confluence of the King and Chalkeetna Rivers. The Chickaloon River trail leads to sites for thundereggs, agatized wood, agates and jaspers.

MEXICO

Mexican residents are beginning to realize that economic giants in the form of geodes and thundereggs lie sleeping beneath the barren hills and deserts of Mexico. It seems that at least one new "geode" discovery is made every year. However, facts concerning the discoveries are often shrouded in legend and fabrication so that it is impossible to cite the date and exact location of a new discovery. However, we will attempt to document and summarize those geodal thunderegg deposits known to date and look forward to documenting future discoveries as they occur.

"Galaxy Geodes"

"Galaxy Geode" is a trade name given to a Mexican thunderegg which has a brown-gray rhyolite exterior with an internal coating of drusy chalcedony. Galaxy geodes are simply hollow thundereggs. They range in size from one to four inches in diameter and were first discovered in 1992. They are found at an undisclosed locality near the Chihuahua-Sonora state line in Mexico.

"Ocotillo Geodes"

Ocotillo is a commercial name given to hollow thundereggs first discovered in 1988, approximately sixty miles southwest of Cuauhtemoc near San Juanito, Chih., Mexico. They have an outer rim and exterior of highly altered pinkish-brown

New Galaxy geode. (Photo by Donnette Wagner; Brad L. Cross Collection)

Ocotillo geode. (Photo by Donnette Wagner; Brad L. Cross Collection)

Palomas geode. (PHOTO BY DONNETTE WAGNER; BRAD L. CROSS COLLECTION)

Red Skin or New Chalce geode. (PHOTO BY DONNETTE WAGNER; BRAD L. CROSS COLLECTION)

San Juan geode. (Photo by Donnette Wagner; Brad L. Cross Collection)

Tamaka geode. (Photo by Donnette Wagner; Brad L. Cross Collection)

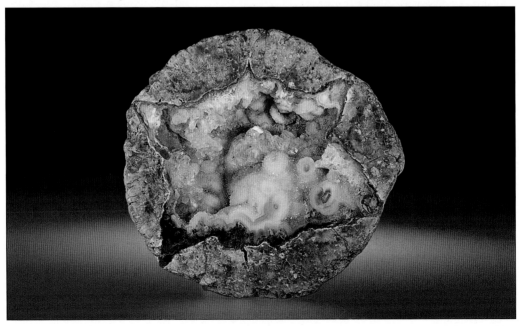

Zacatecas geode. (PHOTO BY DONNETTE WAGNER; BRAD L. CROSS COLLECTION)

rhyolite. Ninety percent of the geode is hollow, providing for a very thin shell. The interiors are lined with clear drusy quartz. Thin, dark striations are abundant in spiderweb-like patterns on the exterior and may be traced inwardly through the rim to the quartz lining. The geodes average three to five inches in diameter, and it is not uncommon to find sparkling double terminated quartz crystals scattered throughout the interior. Many times, the sparkling quartz is surrounded by dazzling calcite crystals.

Unlike typical thundereggs, the interiors of these specimens do not usually exhibit star-like shapes. Occasionally, fragmented pieces of the pinkish-brown rhyolite may be found included in the outermost perimeter of the quartz lining, making them unusual specimens for study.

"Palomas Geodes"

Palomas is a commercial name given to geode-like thundereggs first discovered in 1992. The location is reported to be west of Ciudad Chihuahua, Mexico near Cuahtemoc. A specific location could not be obtained. Palomas geodes have a brick-red rhyolite exterior with an interior lining of clear quartz and sparkling white drusy chalcedony. The thundereggs range from one to four inches in diameter.

"Red Skin or New Chalce Geodes"

These brilliant hollow thundereggs are commercially known as "Red Skin" or "New Chalce Geodes" because of their drusy quartz and botryoidal chalcedony (thus the name "chalce") interior. These geodes are a delight to fluorescent mineral collectors due to their bright green colors under short wave ultraviolet light. Identifying these gems is easily accomplished due to their brick red rim and contrasting white crystal centers. The typical specimens display randomly formed four- to six-pointed star-shaped interiors. The outer surface of the spheres is pale brick-red and does not exhibit elevated exterior ridges typically seen in thundereggs. The average "Red Skin Geode" is relatively small (one and one-half to three inches in diameter). This geode deposit was discovered in 1987 near Cumbres de Majalca, approximately twenty-five miles northwest of Chihuahua City, Mexico.

"San Juan Geodes"

It seems that some new geode or thunderegg is always being found in Mexico. The San Juan geode is no exception. This thunderegg is one of Mexico's newest stars, being discovered in 2001.

As with most all thundereggs, the exterior of the nodule is rhyolite, but these shells are light brown in color with darker brown to earthy red "warts" scattered over the surface. Very thin ridges are exhibited on the outer surface as well. The geodes average two to four inches in diameter. Internally, a four- to five-pointed chalcedony star can be found. A white drusy chalcedony lines the hollow specimens.

"Tamaka Geodes"

The interior of these hollow thundereggs consists of four- to six-pointed star-shaped patterns. Over half of the specimens are hollow with treasures of clear quartz and occasionally white banded chalcedony.

The outer rim consists of a light brown silicified rhyolite, is extremely weathered, and does not typically exhibit ridges or rounded wart-like colloform surfaces. The geodes average three to five inches in diameter. Tamakas were first discovered in 1980 near Palomas, Chih., Mexico.

"Zacatecas Geodes"

These hollow thundereggs contain quartz crystal centers and rhyolite rims of light brown to pink color. Feldspar phenocrysts are commonly seen sprinkled throughout the rhyolite matrix which surrounds the core. Freshly exposed outer surfaces are dark chocolate brown to light lime green and are quite rough with no specific wart-like appearance or ridges as is typical in many thundereggs. Approximately ten percent of these thundereggs contain amethyst. However, upon exposure to sunlight, the amethyst color fades quickly. The vast majority of the interiors contain clear to smoky color quartz. The luster on most of the quartz crystals is dulled by a very thin coating of chalcedonic silica. The geodes average three to six inches in diameter and can exceed twenty-four inches on occasion. The locality is about thirty miles northwest of Ojo Calientes, Zacatecas, Mexico. Workers from the local Ejido Palmillas work various claims in the geode-producing area.

Blue Chalcedony

Found just east of Cerro El Chile on Rancho del Palmar in Northern Chihuahua, this bright and beautiful gem quality blue chalcedony was originally discovered in 1953 by the father and son team of Matilde and Beto Vasquez from Ciudad Juárez, Chih.

Rarely mined as complete thundereggs, the bright blue agate cores were surface collected and most were simply tumble polished by U.S. lapidaries. After the mid-1960s, the location appeared to be forgotten and remained so until 1997 when Ramon Olivas of Juárez and Tony Worth of Pendleton, Oregon, filed claim on the deposit. The gem material is currently being mined as small cores, complete thundereggs, and thin agate veins. The chalcedony ranges from blue-gray to an intense sky blue with little or no "veiling" obvious, although many pieces may possibly have "clouds" in the interior portion. This agate has obviously been subjected to the intense desert heat and the ultraviolet waves of the Chihuahua sun, yet a strong blue color is maintained.

The encompassing rhyolite on the thunderegg is white to light beige in color with an occasional wart-like appearance. These thundereggs range in size from half an inch upwards to three inches in diameter and are considered an "old and true classic" in the Mexican mineral world.

CANADA

British Columbia

Fred Frese and his son built a hunting and fishing lodge in a remote area of British Columbia. To get to the lodge in the high and rough forested country near the Fraser River the 26-mile road near Black Dome, an extinct volcano, turned out to be a true "Gem Trail." After blasting through enormous deposits of variegated obsidian they reached a bed of tons and tons of agate-filled thundereggs. The eggs show a multitude of colors and patterns. Some of the chalcedony centers are clear with white patterns. Some have carnelian or sardonyx centers. Amber or pale yellow T-eggs are numerous, while several shades of green are more unusual. The greens vary from almost peridot color to a light chrysoprase hue. Lots of the T-eggs have crystal centers or a combination of agate and crystal.

Most of the patterns are banded agate and the best of these are vibrant fortification agate. There are also onyx bandings in the T-eggs. The rhyolite "picture frames" of the thunderegg nodules are sometimes tan or beige and sometimes reddish brown or yellow brown. A small proportion are only an inch or so in diameter, quite a few are seven or eight inches in diameter, with an average size of around five inches in diameter.

Agates and jaspers and other gem materials abound in this area. The most beautiful jasper is blue in color and orbicular in design.

Thundereggs have also been found northeast of Kamloops, south of Vanderhoof, and near Clinton as well as several other places. Many areas of northern British Columbia are mountainous and cut by large rivers. The farther north one goes the fewer roads and villages are to be found. There are several Canadian National Parks in spectacular wilderness country.

• • •

The *Dictionary of Geological Terms* defines thunderegg as a "geode-like body commonly containing opal, agate or chalcedony, weathered out of welded tuff or lava." John Sinkankas points out that they differ from ordinary geodes in their mode of formation and internal structure and in the geologic setting in which they occur. Most books explain and describe geodes and thundereggs in sequence in that order and with emphasis on geodes. Most amateurs refer to hollow crystal-lined thundereggs as geodes. Probably the time will come when they will be recognized not as first cousins but as separate families.

OPENING THUNDEREGGS

Most nodules and geodes have external features that an experienced lapidary artisan can identify by sight for cutting so as to capture all of the best features within. For example, the external raised ridges on a thunderegg give an indication of its internal structure. A thunderegg with a single, central ridge circumscribing its equator indicates a single lensoidal interior; whereas nodules with multiple external ridges suggest a more three-dimensional center that may appear as a geometrical star when cut in appropriate orientation.

Thundereggs need to be cut with a diamond saw to reveal the internal core. Anyone who has cut a number of these nodules has developed a pet theory as to proper orientation. Each theory is usually based on the pattern of the external ridges, revealing a geometric star. Regardless of the method, it remains a matter of luck to find a *perfect* star in the first place. One commonality that all theories share is that the cut should be made through the long axis (for the widest face) of the nodule. This particular orientation will yield a cut face showing the sequences of agate, flat layered "waterline" bands, and crystalline formations "right side up."

Many thundereggs are very ordinary and are not showy. Do not be discouraged if you cut a half dozen eggs or more to find a beauty. Do not use a hammer in the field or at home. Thundereggs are always sawed, not broken in half like Keokuk geodes. However, do examine the first one you cut before cutting the others to check your technique and to gain insight about the internal pattern.

Unlike concretions, septarians have hollow sections with crystals, fossils, or both. Utah's are lapidary quality. 7 inches. (ZEITNER SPECIMEN; PHOTO BY DAVID PHELPS)

CHAPTER SIXTEEN

Almost Geodes: Concretions and Septarians

— JUNE CULP ZEITNER —

— BRAD L. CROSS —

Concretions are hard, compact accumulations of mineral matter that grow within sedimentary rocks such as sandstone and shale. The mineral matter concentrates locally in the host rock, cementing it together to form harder zones or nodules. Upon opening a concretion, one usually finds that the cementing material precipitates locally around a nucleus, often organic, such as a piece of shell or fossil, leaf, tooth, or sometimes around a mineral crystal or sand grain. The word "concretion" is derived from the Latin "con"—meaning "together"—and "cresco"—meaning "to grow."

The odd shapes of concretions arouse curiosity and they can often be mistaken for fossils, bones, meteorites, or other unusual objects. Concretions are commonly sub-spherical, but they also can have shapes like boxes, blocks, flat disks, canonballs, or even resemble parts of a human body such as a foot or rib. It is sometimes hard to believe that they formed by natural means.

Record size concretions of Rock City, Ottawa County, Kansas cover a huge area. The largest are over twenty feet in diameter.

Concretions vary in size, shape, hardness, and colors. They can be objects that require a magnifying lens to be clearly visible to huge bodies tens of feet in diameter and weighing several hundred pounds. They are commonly composed of a carbonate mineral such as calcite, but sometimes an iron oxide or hydroxide such as goethite, or sometimes an amorphous or microcrystalline form of silica about a nucleus. They can also be composed of other sedimentary minerals that

235

The giant Rock City concretions of Kansas are a popular landmark.
(Photo Courtesy Kansas Geological Survey)

Concretions can be fun. (Photo by James Boles)

include dolomite, siderite, pyrite, barite, and gypsum. Concretions are usually very noticeable features, because they have a strikingly different color and hardness from the surrounding rock.

Some concretions may be hollow, the center being empty or filled with loose powdery clay or sand, or a detached hard lump resembling a nut. Concretions appear in nodular patches, concentrated along bedding planes, protruding from weathered cliff sides, randomly distributed over mud hills, or perched on soft pedestals.

Most concretions form as ground water with dissolved iron, silicon, calcium, or other chemicals will "drop" the mineral as iron oxide, calcium carbonate, or silica solids when chemical conditions change, adding them a little at a time as a thin layer. Many such layers build up, having different concentrations of the compounds, and sometimes showing different colors.

Recently hydrologists have become interested in elongate concretions. These specimens range in size from a pencil on up to some that resemble fallen logs. They are thought to form from flowing ground water with the long axis of the concretion oriented parallel to the ground-water flow direction. Measuring the concretion orientation can provide a direct measurement of the past ground water flow orientation over a large area.

Peter Mozley with the Department of Earth and Environmental Science at New Mexico Tech reports some of the most unusual concretion nuclei are found in a modern salt marsh in England. Siderite (an iron carbonate) concretions in the marsh formed around World War II era military shells, bombs, and associated shrapnel. The concretions formed preferentially around the military debris because it provided an abundant source of iron for the siderite.

AMMOLITE: CANADA'S FOSSIL GEM

The fiery iridescent shell of fossil ammonites found in the Bearpaw formation of Alberta, Canada, resemble precious opal in many ways. The play of color of opal, ammolite, fire agate, and iris obsidian is due to their responses to lightwaves. Dr. Frederick Pough suggested calling such materials *spectrochromatic.*

The ammonites are entombed in heavy dark concretions that must be removed before the colorful gem material, iridescent aragonite, can be seen. Some concretions, broken by nature, are scattered over the fossiliferous acres.

A gem-covered ammonite from an Alberta, Canada concretion is paved with cuttable ammolite, reminiscent of black opal. 12¼ inches.
(PHOTO COURTESY OF KORITE INTERNATIONAL, INC.)

An ammolite necklace by Korite jewelers has a vivid 1-inch cabochon set in gold. Zeitner necklace.
(PHOTO BY DAVID PHELPS)

238

The concretions are not the same as septaria. There are no mazes of open cavities. True concretions are best described by John Sinkankas's definition: "rudely spherical masses formed by deposits around a nucleus within enclosing rock and not a cavity."

Carbonate cemented shale concretions can have baculites or other marine fossils inside, or ammonites, or they could contain nothing at all, in which case they are referred to as mudballs. The size range of the concretions is from three-quarters of an inch to well over six feet in diameter. The concretions are mined near the city of Lethbridge. Many are more than 300 feet below ground level.

Aragonite with vivid iridescent luster from the area is thicker and harder than other deposits. All the colors of the spectrum appear and interesting patterns add to the unique fossil gem. I like the geometric brecciated patterns best, but those resembling cobblestones, flowers, ribbons and feathers are beautiful variations.

"Ammolite" is used for jewelry, sometimes elegant, pricey, jewelry set in gold and accented with diamonds, rubies, or emeralds. For jewelry the ammolite is usually made into doublets which are capped with clear quartz to make it more durable, or fashioned into triplets with the black stone backing, giving the finished cabochon a "black opal" appearance.

More Concretions

Concretions are interesting and variable in size and shape, but not rare. They have occurred in most states. Many are still being discovered. Some are forming even now.

Hall Summit in Louisiana's Sabine Parish has abundant ironstone concretions. Ben Schaub reported glacial clay concretions from the Northhampton, Massachusetts area. Most are round and some are only rings, like stone bracelets.

Along the Scituate River in Connecticut strange and irregulary shaped clay concretions occur. These are rod shaped, worm shaped, even foot shaped.

Near Riverton, West Virginia, concretions with barite linings are found.

Clay concretions in Wyoming, Nebraska, Texas and Arkansas resemble primitive models of animals, plants, or manufactured objects, or inanimate shapes, all with hidden beauty or wonder inside.

Tiny to gigantic freeform shaped concretions occur in northern Ontario, Canada, near Abitibi.

Perfectly round concretions found in western North Dakota along the Cannonball River, called "cannonballs" by the residents, were once abundant. During World War II, I knew a lady who outlined her flower garden with these balls of stone. One day she found them all missing and immediately called the police. "Someone stole my cannonballs." She soon had a visit from high-ranking officers!

UTAH'S GEM CONCRETIONS

A beautiful American gem occurs as concretions. It is variscite, a phosphate sometimes compared to turquoise. Variscite has been mined in several Utah locations but to me the most desirable was mined near Fairfield. Some of the spectacular nodules rival most gems with hardness lower than quartz.

With greens, blue greens, yellow greens, mixed with other phosphates in hues of yellow, orange, gray, rust, ivory, and lavender, there are patterns most designers would envy. Spheres, slabs, cabs, inlay, and bowls have delighted most collectors. Now hard to get, the best quality is among the most pricey American gems.

Some of the mineral-providing colors other than the greens are yellow crandallite, reddish strengite, smoky gray gordonite, lavender deltaite, and red-orange hydroxlapatite. Some of the Fairfield nodules were large enough to make many slabs up to twenty inches or more in diameter, each slab a work of art.

Variscite is rather porous and heat sensitive. Only 4.5 in hardness, it has a toughness which helps its cutability. Seldom seen at shows after the mid-1950s, the Fairfield concretions were replaced for a while with less colorful and smaller nodules from Toole, Utah.

A favorite of mineralogist-author-lapidary John Sinkankas, variscite is exquisitely depicted in one of his original watercolor paintings for *Gemstones of North America*.

ILLINOIS' FOSSIL CONCRETIONS

Typical of smaller concretions are the so called "fern fossils" of Mazon Creek in northern Illinois. Hundreds of species of Pennsylvanian flora and fauna have come to light when the concretions are opened to reveal a perfect mold-and-cast

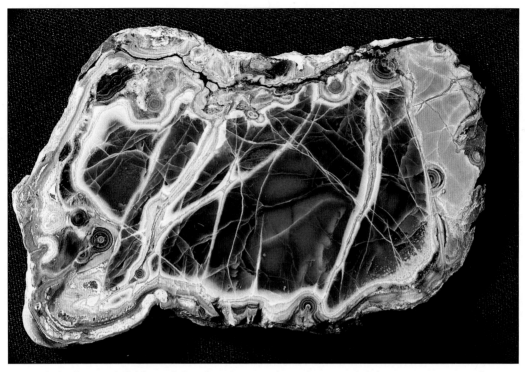

Variscite from Fairfield, Utah is a favorite American gem material that occurs as concretions.
(COURTESY LIZZADRO MUSEUM OF LAPIDARY ART; PHOTO BY RUSSELL KEMP)

These collectors are looking for "fern fossils" in a coal mining site near Morris in northern Illinois. (ESCONI PHOTO)

Concretions from Mazon Creek, Illinois, hold an incredible history of the flora and fauna of the Pennsylvanian. Pectoperos vestita. 9 inches. (Zeitner Specimen; Photo by David Phelps)

Fossil-bearing concretions are numerous in Illinois coal seams. Split nodules often reveal compressions of seed ferns. 6 inches. (Zeitner Specimen; Photo by David Phelps)

picture of life 300 million years ago. Shale exposures in the coal mines of northern Illinois have yielded literally tons of the concretions which average between three and six inches in length. Some are much smaller, for example those carrying seeds, while others, perhaps holding a large frond, may be eleven inches or more. The concretions are cemented by calcite or siderite. Those including siderite, an iron carbonate, are reddish on the exterior, others are gray. The Field Museum has an exceptional collection of these important concretions.

Concretions are a close cousin of geodes.
(PHOTO COURTESY MISSOURI DEPARTMENT OF NATURAL RESOURCES, GEOLOGICAL SURVEY AND RESOURCE ASSESSMENT DIVISION)

UTAH'S SEPTARIAN NODULES

Septarians are a form of concretion and occur in various locations throughout the world, but none is comparable to the wonderful nodules found in Utah. These beautiful crystal-filled specimens are dug from the hills of southwest Utah. The most productive concentrations are found near the town of Orderville, but because of their value most of the area is protected by private claims. You can dig on some of these claims if you check with the owners beforehand. Be warned though, Septarians are generally found many feet below the surface.

These Utah beauties started their formation some 100 plus million years ago when the Gulf of Mexico reached what is now Southern Utah. Decomposing sea life buried in the mud at the bottom of the ancient sea had a chemical attraction for the sediment around them, forming "mud balls." As the land continued to rise and the sea drained away, the concretions dried and shrank. The shrinkage caused a network of wedge-shaped cracks to form that were later filled with calcium carbonate solutions, forming sparkling yellow calcite crystals. A thin wall of calcite was transformed into aragonite (the brown lines) dividing the outer gray clay exteriors from the calcite centers. Because of this dividing wall (*septum* in Latin), the nodules are called Septarians.

These cracks that are wider toward the center and die out as they near the outer edge are crossed by a series of smaller cracks that are roughly parallel to the outside of the nodule. It is believed that since the outside of the mud ball had already dried and hardened, the shrinkage was confined to the center of the nodule.

Septarian nodules were originally found on the surface of the ground, weathering out of layers of shale and clay cliffs along Muddy Creek near Orderville. Today, heavy equipment is used to dig large open pits. Using a big tractor with a blade in the front and ripper teeth behind, the hillsides are terraced. A great amount of earth is moved until a few broken concretions appear on the surface. Then, using the teeth on the back of the tractor, several feet of the soil are carefully loosened. It may take several attempts to break up the clods and separate the nodules from the hard clay. After the soil is pretty well pulverized the tractor blade is lowered and moving along slowly, several inches of earth are pushed aside. As the soil spills around the ends of the blade, the

Septarian Nodule, Orderville, Utah.
(PHOTO BY DONNETTE WAGNER; BRAD L. CROSS COLLECTION)

nodules roll out like potatoes and are picked up by workers. Ron Weaver, a septarian miner, says the best quality septarians are found twenty to thirty feet deep. Freezing and thawing left many specimens from higher levels fractured or broken.

Anywhere between two and six feet below the layer containing septarians, the flattened spiral shells of an ancient cephalopod, the ammonite, are found. These are quite a treasure in any collection.

Radial and tangential ridges distinguish septarian spheres from many shapes.
(PHOTO KANSAS GEOLOGICAL SURVEY)

SOUTH DAKOTA'S SEPTARIA

The bonanza Pierre Shale "concretions" of South Dakota (actually septarians) are rated highly by both crystal collectors and fossil collectors. The huge septaria occur in western South Dakota. The best exposures are along Elk Creek in Meade County. Prismatic barite crystals, transparent and lustrous, are the big prizes in the septaria. Called "golden barite," the crystals are surrounded by excellent yellow calcite crystals covering the numerous walls of the septarium. Many of the barite crystals are usually from two to four inches long. They are not in clusters,

246

but often several of them are perched on a small section of the concretion. The best one we found was a 5-inch single, bright, transparent, and glowing in a field of sunny calcite. Willard Roberts in *Mineralogy of the Black Hills* tells of a record giant 12 inches long. A particularly fine specimen from Elk Creek is in the Smithsonian. Roberts introduced the glamour barites to the mineral enthusiasts at the California Federation Show in 1951.

Albert and I started digging out and breaking up the septaria in the early 1950s. It was not hard to break the weak tops with a maul, but the work barely began there. The barite crystals could be anywhere in the maze of separate sections, and their perches were extremely fragile because of the perfect basal cleavage. One false move and a shining crystal would tumble through the channels to the bottom, knocking off many more on its journey. If the septarium looked like a good one when opened, the only thing to do was to proceed very slowly taking the septarium apart piece by piece examining each one carefully and placing

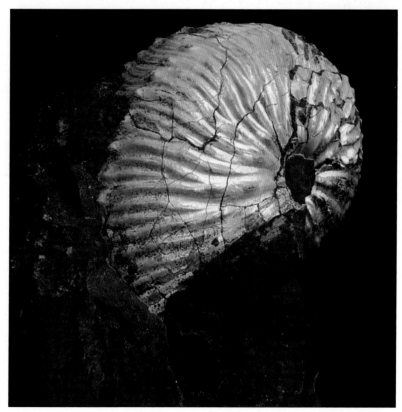

South Dakota Pierre Shale septarians often contain Cretaceous cephalopods. The iridescent scaphite is partially embedded in a septarian concretion. 7 inches.
(SPECIMEN AND PHOTO BY DAVID PHELPS)

Golden barite crystals are prized specimens from South Dakota's huge septarians. 4 × 6 inches. (PHOTO BY DAVID PHELPS; DAVID PHELPS COLLECTION)

each part with the barite crystals in a separate box of the right size. We always stuffed clean rags around the base of each specimen so it would not move. One of our best is in the Black Hills Museum of Natural Science.

Some of the septaria harbored the most beautiful ammonites in the world, often with iridescent rainbow colors, looking like a superb opal carving. Sometimes a large one will hold only one king size specimen, and other times there will be a whole family of cephalopods, large, complete, and showy. John Sinkankas, author, lecturer, artist, and mentor to many lapidaries considered the South Dakota barite as the best for faceters of rare cutting material who wanted to add barite to their list.

CONCRETIONS OF OKLAHOMA

Cimmaron County, Oklahoma has cherty limestone concretions, smooth on the surface and dull in color. The patterns are typical of septarians. The shapes are rather round and regular. Lapidary hobbyists have found them to be polishable and suitable for decorative purposes. However, some other attractive Oklahoma concretions are agatized.

The silicified Kenton concretions are unique and highly prized. Eerily irregular in shape with edges scalloped and serrated, the agatized interiors are crowded with ancient but ultra-modern designer patterns created by algae.

When cut open the patterns are surrounded by graceful borders sometimes resembling banded agate. When examined with a ten-power lens each slab will show hundreds of shapes each touching several more. There are rods, cones, tubes, ellipses, circles, eyes, teardrops, leaves, and chisels, all in hodge-podge arrangement. Reefs of primeval algae called stromatolites are found in New York, Michigan,Wyoming, Montana, Ontario, Canada, and other places; but agatized concretions such as these from Oklahoma are rare. To me they are exceptionally interesting, both for their science and their art.

Classed as primitive plants, blue-green algae dates from shallow Precambrian seas where they thrived with red algae, green algae, and other organisms. The algaes are still with us!

Along with these fantastic agatized algal concretions from the Cimarron River highlands are nodules of silicified sponges and corals. A popular collecting area is near Black Mesa, the state's highest point. The late Marie Kennedy had a fine collection of Oklahoma concretions. Another lucky collector in this rugged, semi-arid region near the New Mexico border was photographer Jim Slack of Tulsa. Eroded badlands panoramas are distinguished by spires, columns, giant sandstone "mushrooms," and even "faces."

OTHER SEPTARIA

Similar fossil bearing septaria occur in the Bearpaw formation of Montana. The Bearpaw is the equivalent of the Pierre Shale. The ammonites and other cephalopods accompanied by calcite crystals are often huge and lustrous.

Kansas has Pierre shale fossils and crystals in septaria. Septarian concretions are found in several Midwest states. One of the better Kansas locations is along

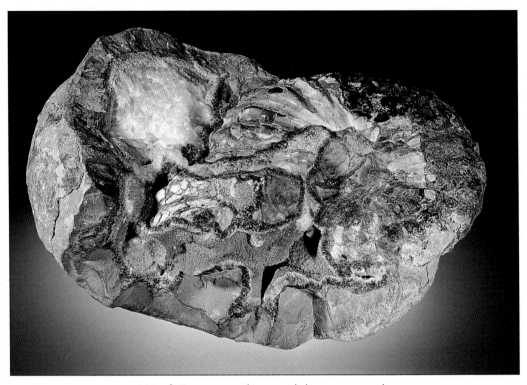

A North Texas concretion containing an ammonite.
(Photo by Donnette Wagner; John Tutor Collection)

Eagle Tail Creek near Sharon Springs bordering Wallace and Logan Counties. The septaria of Osborne County has attractive yellow calcite crystals.

Ohio has attractive septarian concretions in Lorain County along the Vermillion River. They are also found in Erie County. A good exposure is in Mill Hollow – Bacon Woods State Park where there is also a nature museum and picnic area.

Septaria up to eight feet containing pieces of petrified wood or parts of fossil sharks are quite obvious along the Huron River. The area is accessed by Lamereax Road north of Morceville in Huron County.

Ohio has plentiful ironstone concretions of highly variable shapes and sizes in Wayne, Jefferson, Carroll, and Athens Counties. The septaria contain siderite, calcite, quartz, barite, and sphalerite. The barite and calcite are both white. Fossils also occur in these concretions.

A location for siderite concretions is in the vicinity of Shelecta, northwest of Indiana, Pennsylvania near Crooked Creek. The concretions, which have barite crystals, are famous for yielding western Pennsylvania's first wurtzite crystals. This formation, Glenshaw, is also fossiliferous.

Near Altoona a road cut and wash expose large concretions up to several feet in diameter. Crystals of calcite, barite, dolomite, and pyrite and rarely the zeolite phillipsite adorn these concretions. The nearest town is Grazierville, a small town on State 220 just south of Tyrone and north of the town of *Culp* (of all places!)

Iowa has a pastel pinkish barite east of Bussey as well as other sites in Marion County. Marion and Mahaska Counties also have colorless and white barite and white and ivory colored calcite in septaria, sometimes with fossils.

North Dakota's Pierre shale septaria are found in Emmons County along Beaver Creek, which flows into the big Missouri near Lake Oahe.

Washington County, Arkansas, septarian concretions have white or clear calcite crystals and marine fossils. Pyrite is also present, sometimes replacing fossils similar to those found in the Centennial quarry of Sylvania, Ohio.

Texas has a notable site for large fine septaria. During a road project in Tarrant and Dallas Counties enormous septaria were collected with ammonites weighing up to 100 pounds. Some of the septaria reached lengths of up to fourteen feet! The crystals were calcite and aragonite. The calcite in these septarian concretions is yellow and orange and is usually brightly fluorescent. The preservation in many of the septariums is excellent. The construction of Loop 12 around Dallas (particularly near the intersection of Interstate Highway 30) through the Cretaceous Eagle Ford shale was a prolific source of crystals and invertebrate marine fossils.

SUMMARY

*S*o after examining hundreds of sedimentary geodes from multiple sites what conclusions are reached? Geodes are much more varied and interesting than one might imagine. A large proportion is not even round as old definitions insist. Many are of fossil origin. Most show progressive inward growth toward a center which may be hollow or packed with crystals. In size they range from peas to winning pumpkins at state fairs. Colors cover the spectrum with shades and tints of rainbow hues. They are prime examples of the old saying "no two are alike." Although primarily chalcedony and quartz, some are barite, celestite, calcite, or goethite. It would not surprise me to find geodes revealing new inclusions, replicating mysterious shapes or composed of rare minerals.

The stunning features of geodes born in volcanic rocks also provide snapshots of the earth's past. Great explosive eruptions and profound collapses of the ground, enormously thick lava flows, uplift and extensive faulting, mobile fluids migrating through seemingly solid rock, and the erosive power of flowing water can all be seen in many volcanic geodes and the localities in which they are found. The geodes tell a remarkable story of the earth. And what variation! Colorless eye-clear quartz to exotic rare minerals perched on top of dark purple amethyst. The experience of holding a small geode in your hand and being the first to marvel at its complex nature-sculpted crystals is every bit as exciting as stepping foot inside a giant Brazilian amethyst geode. New discoveries are made every year and the best may be yet to come.

We have two favorites. The beauty we are looking at, and the spectacular one we dream of finding.

GEOLOGIC TIME SCALE

Time Division Name			Time Span (mya = million years ago)	Geode/Host Rock
CENOZOIC ERA	QUATERNARY PERIOD	Holocene Epoch	0.01 mya to Present	
		Pleistocene Epoch	1.8 to 0.01 mya	Claystone Concretions
	TERTIARY PERIOD	Pliocene Epoch	5.3 to 1.8 mya	Agatized Coral Geodes (Florida); Oregon Thundereggs
		Miocene Epoch	23.7 to 5.3 mya	
		Oligocene Epoch	36.6 to 23.7 mya	
		Eocene Epoch	57.8 to 36.6 mya	Mexican "Coconuts"; Other Mexican Geodes & Thundereggs
		Paleocene Epoch	65 to 57.8 mya	
MESOZOIC ERA	CRETACEOUS PERIOD		146 to 65 mya	Septarian Nodules (Utah); South Dakota Golden Barite Concretions
	JURASSIC PERIOD		208 to 146 mya	Brazilian Geodes; Uruguayan Geodes
	TRIASSIC PERIOD		245 to 208 mya	
PALEOZOIC ERA	PERMIAN PERIOD		286 to 245 mya	Nebraska & Kansas Geodes
	PENNSYLVANIAN PERIOD		325 to 286 mya	Illinois "Fern Fossil" Concretions
	MISSISSIPPIAN PERIOD		360 to 325 mya	Indiana Geodes; Keokuk Geodes
	DEVONIAN PERIOD		410 to 360 mya	
	SILURIAN PERIOD		440 to 410 mya	
	ORDOVICIAN PERIOD		505 to 440 mya	
	CAMBRIAN PERIOD		544 to 505 mya	
PRECAMBRIAN ERA	PROTEROZOIC EON		2,500 to 544 mya	
	ARCHAEAN EON		3,800 to 2,500 mya	Blue-green Algae Concretions (Oklahoma)
	HADEAN EON		4,600 to 3,800 mya	

FREQUENTLY ASKED QUESTIONS
ABOUT GEODES

Q: Are there clues on the outside of a geode as to what will be found inside?

A: No. There are no features, characteristics, specific shapes, colors, or markings that will provide even a clue as to what will be found within the interior of a geode.

Q: What chemical gives smoky quartz its color?

A: The color of smoky quartz is due to what is called a color center. Color centers arise when an electron becomes trapped in a vacancy in a crystal's structure and vibrates at a certain frequency dictated by the size and type of "hole" or vacancy where it is trapped. The vibrating electron absorbs certain frequencies from white light and the remaining frequencies are the color that our eyes perceive.

These "holes" open up when a few atoms of some impurity – an element that is not part of that minerals essential structure – become incorporated into what is called the crystal lattice. Radiation provides the energy necessary to knock an electron off an atom and send it careening about until it gets trapped in one of these holes in a manner similar to the way a well-executed break sends one or two balls into the pocket of a pool table. As long as the electron is vibrating in its hole, it will produce color. If, however, the crystal is warmed, the electron can pick up that heat and use the energy to jump out of its hole – thus ending its "colorful" career.

The smoky color center is a relatively "shallow" hole. As a result of these shallow color centers, the color in smoky quartz (as are those in a number of other gems) is not very stable and can fade when exposed to even modest heat.

In smoky quartz, the impurity atom is aluminum that has usurped a place on the lattice that rightfully belongs to a silicon atom. It takes the most incredibly small amount of aluminum to do the trick – something on the order of 0.002 percent by weight.

Smoky quartz is colored by a color center and therefore will fade if heated. Likewise, after bleaching by heat, the color can be restored by another dose of radiation, natural or artificial.

Q: What element or chemical gives amethyst its color?

A: Like smoky quartz, the color in amethyst is due to a color center. However, the impurity atom is iron, which upon being irradiated provides the lovely lavender to purple color.

Q: How can I tell the difference between a Brazilian geode and a Uruguayan geode?

A: Rarely do you find a completely intact geode from Uruguay. When you do, it will usually have an excessive amount of basalt surrounding the geode shell. The Uruguayan geode may contain a darker color amethyst. Uruguayan amethyst crystals are also commonly smaller than those from Brazil. Most Uruguayan crystals are not larger than 1 cm. Those more than 3 cm are considered large. Unlike Brazilian geodes, most Uruguayan material is backed by a blue-gray banded agate and the specimens are, on average, substantially thinner. Finally, most Uruguayan amethyst consists of small angular plates of tightly interlocking crystal points.

Q: I have a large, tall amethyst geode. What are the factors in determining its price?

A: First, you need to determine the origin of the geode. It may come from Rio Grande do Sul, Brazil, or from near Artigas, Uruguay, on the Brazil/Uruguay Border. If the crystals are smaller (5 to 15 mm), it is likely Uruguay. If the run of the crystals are larger (10 to 30 mm), the likely source is Rio Grande do Sul. Brazilian amethyst tends to be lighter in color and tends to exhibit zoning near the crystal tips.

With respect to value, two situations are presented. The geode may be kept for a specimen or for a source of lapidary material. The shade and clarity of the amethyst will help determine the choice. Medium or light color material should probably be kept as a specimen. A geode which has dark color and is relatively free of flaws may be worth more as facet rough. You must ask the question, "Does the amount of faceting material surpass the value that it might have as a specimen?"

In 2004, specimen material sold in bulk averages between $10 to $15 per kilo. Single fine collection pieces are sold by the "how pretty they are" method, but generally average $45 to $50 per kilo. Large and dark crystals are usually cobbed and sold as gem rough. Cobbed dark faceting rough sells on average for $500 per kilo. Extra dark color amethyst can sell for up to $10,000 per kilo.

If the geode is kept as a specimen, an additional factor in determining its price is its weight vs. its dimensions. A geode with very thick sidewalls or an extra heavy base is not as desirable. These thick walls and base decrease the area or the "window" of view and add nothing to its beauty. One additional factor to consider is the symmetry or outside shape of the geode. The more symmetrical geodes command the best prices.

Q: Is it possible that my large citrine geode is actually heat-treated amethyst?

A: Yes. The crystals in your geode started out as amethyst, but somewhere between the mine and the rock shop, they got "cooked." Much of the Rio Grande do Sul amethyst is heated in the 800° F range for about an hour. All of the amethyst from this state treats to a light to dark red/orange. Specific locations produce a higher percentage red/orange than others do.

Q: How can a Florida geode have one pocket with crystals and another with botryoidal chalcedony?

A: The crystal-lined pocket is entirely closed, while the botryoidal pocket had an opening where mineral laden waters could leave new inclusions or change the colors, shapes, or other physical attributes.

Q: Are the geodized fossils of the Salem-Warsaw-Keokuk formation unique?

A: Perhaps not. Such fossils are found in other parts of the world. Geode puzzles have not been completely solved. Perhaps many other geodes had marine life "seeds" and their growth patterns obscured their origins. "Puffed geodes" are an example. Only long intensive study can reveal their true history.

Q: How do you determine the value of a geode?

A: Does it have unusual or rare beauty, inclusions, colors, or size? Is it from a locality which yielded very few geodes? Does it fill a gap in the scientific study of geodes? Was it in the collection of a famous mineralogist? *Has someone offered you a price you can't refuse?*

Q: Why is the Mississippian Period so important for the sedimentary geodes?

A: In the Mississippian Period, which lasted 30,000,000 years, shallow seas were spread widely over much of the interior of North America. Much of the East Coast was swampy lowland. Sea sediments were rich in anhydrite, calcite, and other minerals needed for geode formation, as well as flora and fauna remnants that were the seeds for geodes to form in the seabeds and continue their growth in brackish sediments.

Q: Where can I find geodes?

A: Geodes are found throughout the world, and can possibly be found wherever you find igneous or sedimentary rocks. However, the deposits are isolated and are not plentiful. There are several commercial geode mines where you can harvest your own treasures. Examples include mines south of Keokuk, Iowa; St. Francisville, Missouri; and Southern Utah near Zion National Park.

There are also excellent collecting sites along road cuts or highway excavations. Examples include Highway 136 south of Keokuk, Iowa, and Country Road 480 near Warsaw, Illinois. Geodes can also be collected throughout Monroe County, Indiana; around the small town of Loretto in south central Tennessee; the tributaries of the Green River in south central Kentucky; and Dugway, Utah. Don't forget about Rockhound State Park near Deming, New Mexico; the thundereggs are most unique. **Always remember to never trespass and always ask permission ahead of time before entering a property.** Local gem and mineral clubs are often a good source for finding a locality and checking on its collecting status.

Q: How old are geodes?

A: The age of a geode is not easily determined. Let's use a geode found in an igneous rock as an example. The geode is not the direct product of the volcanic

258

activity. The volcanic activity simply provided a hole for the geode to form in. The gas cavity may have sat empty for millions of years before mineralized solutions entered the host rock and deposited the ingredients necessary to form a geode. Just because our host rock is 65 million years old doesn't mean the geode is the same age. All we can say is that the geode is no older than 65 million years. The only way to determine an approximate age of a geode is through complex oxygen isotope studies carried out in research laboratories.

However, we can make some pretty good guesses. Most sedimentary geodes are about 350 million years old, having originated in the Mississippian Period in the Osagean division. The Brazilian/Uruguaian igneous geodes are between 119 and 149 million years old while their "young" relatives, the Mexican "coconuts," are about 44 million years old.

Q: Can you find gold inside geodes?

A: No. Contrary to popular belief, you cannot find gold inside a geode. While gold tends to occur mostly where igneous activity has taken place, the gold usually forms as a result of what geologists term "contact metamorphism." Contact metamorphism involves escaping high-temperature gaseous emanations occurring from hot magma intruding a previously formed rock. Geodes are formed in fairly cool environments and are rarely found in contact metamorphic zones. That is to say, gold usually forms at very hot temperatures while geodes form at much lower temperatures. While you may find minerals that appear as "gold" in a geode, they are likely pyrite or marcasite – common minerals found in many geodes.

Q: How do you know if a geode is hollow?

A: About the only way to determine if a geode is hollow is by "specific gravity." Although minerals seem nearly of equal weight in specimens of the same size, there are sufficient distinctions among them to make their relative weights useful in identification. Relative weights are called *specific gravities* and refer to the weight of a mineral compared to an equal volume of water. The same is true with geodes. Every geode will have an individual specific gravity as compared to another geode of equal size. The only practical way of determining this specific gravity difference is through "hefting" the geode. That is, holding the geode in your hand and testing its weight by lifting it and comparing that weight to another geode of equal size. Take two geodes of equal size, place a geode in the palm of each hand, and compare the weight of one to the other. The geode that feels the lightest is the geode with a larger hollow center.

BIBLIOGRAPHY

Books

Camp & Richardson, 1999, *Roadside Geology of Indiana*, Mountain Press.

Carlson, Ernest, 1991, *Minerals of Ohio*, Ohio Geological Survey.

Cross, Brad L., 1996, *The Agates of Northern Mexico*, Burgess Publishing.

Cross, Brad L., 2001, *Gem Trails of Texas*, Gem Guides Book Company.

Erd & Greenberg, 1960, *Minerals of Indiana*, Indiana Geological Survey.

Helton, Walter, 1964, *Kentucky Rocks and Minerals*, Kentucky Geological Survey.

Macpherson, H.G., 1989, *Agates*, National Museums of Scotland and the British Museum (Natural History).

Mitchell, James, 1991, *Gem Trails of Nevada*, Gem Guides Book Company.

Mitchell, James, 1996, *Gem Trails of Utah*, Gem Guides Book Company.

O'Donaghue, Michael, 1987, *Quartz*, Butterworth and Company.

O Harra, Cleophas, 1920, *The White River Badlands*, South Dakota School of Mines.

Roberts & Rapp, 1965, *Mineralogy of the Black Hills*, South Dakota School of Mines.

Shaub, Benjamin M., 1989, *The Origin of Agates, Thundereggs, and Other Nodular Structures*, The Agate Publishing Company.

Schaub, Wm., 1996, *Fossil and Lapidary Guide to Kentucky*, Schaub.

Sinkankas, John, 1959, *Gemstones of North America, Volume I*, D. Van Nostrand Company, Inc.

Sinkankas, John, 1976, *Gemstones of North America, Volume II*, D. Van Nostrand Company, Inc.

Sinkankas, John, 1997, *Gemstones of North America, Volume III*, Geoscience Press.

Sinotte, Steven, 1969, *The Fabulous Keokuk Geodes*, Wallace.

Sinotte, Steven, 1972, *Geode Country*, Geode Press.

Smith, Edward, 2003, *Geodes of the Midwest*, Secret-of-the Pros Press.

Stepanski, S. and Snow, K., 1996, *Gem Trails of Pennsylvania and New Jersey*, Gem Guides Book Company.

Zeitner, June Culp, 1996, *Gem and Lapidary Materials*, Geoscience Press.

Zeitner, June Culp, 1998, *Midwest Gem, Fossil, and Mineral Trails: Great Lake States*, Gem Guides Book Company.

Zeitner, June Culp, 1998, *Midwest Gem, Fossil, and Mineral Trails: Prairie States*, Gem Guides Book Company.

BIBLIOGRAPHY

MAGAZINES, PERIODICALS, AND PROFESSIONAL PUBLICATIONS

Allaway, Wm., "Indiana Geodes," *Earth Science*, February 1960.

Barwood & Schaffer, "Silicification in Geodes of the Mississippian Sanders Group," Indiana Geological Survey, 2002.

Bassler, Ray, Smithsonian Institution, "The Formation of Geodes with Remarks on the Silicification of Fossils," 1909.

Baxter, H. & S., "Suwanee River Coral," *Gems & Minerals*, March 1979.

Bhaga, D. & Weber, M.E., "Bubbles in Viscous Liquids: Shapes, Wakes, and Velocities," *Journal of Fluid Mechanics*, 105:61-85, 1981.

Borschel, Ken, "Iowa Geode Varieties," *Earth Science*, December 1959.

Borschel, Ken, "More About Iowa Geodes," *Earth Science*, September 1969.

Chown and Elkins, "The Origin of Quartz Geodes Through Silicification of Anhydrite Nodules," *Journal of Sedimentary Petrology*, September 1974.

Carrillo, Hector and Jeannette, Mexican Coconut Geodes, personal communication, 2002.

Cross, Brad L., "The 'Coconut' Geodes of Northern Mexico," *Rock & Gem*, March 2002.

Cross, Brad L., "Coconut Geodes," in Symposium on Agate and Cryptocrystalline Quartz, Program and Abstracts, Golden, Colorado, September 2005.

Cross, W., "Constitution and Origin of Spherulites in Acid Eruptive Rocks," in *Philosophical Society of Washington*, Volume 11, pp. 411-43, 1891.

Dake, Henry, "Petroleum Filled Geodes," *Mineralogist*, 1938.

Dodd, Robert, et al, "Ramp Creek and Harrodsburg Limestones," Indiana University, October 1986.

Douglas, Dale, "Changed and False Fossils," *Earth Science*, March 1960.

Finkelman, Robert, "A Scanning Electron Microscope Study of Minerals from Chihuahua, Mexico," *The Mineralogical Record*, Volume 3, Number 5, 1972.

Finkelman, Robert, "A Guide to Identification of Minerals in Geodes," *The Mineralogical Record*, February 1974.

Finkelman, Robert B., etal, "Manganese Minerals in the Geodes from Chihuahua, Mexico," *Mineralogist Magazine*, Volume 39, 1974.

Fleener, Frank L., and Ben Hur Wilson, "Natures Pandora Boxes," *The Mineralogist*, September 1948.

Frazier, Si and Ann, "Thunder Showers," *Lapidary Journal*, April 1993.

Frazier, Si and Ann, "Shades of Smoky," *Lapidary Journal*, December 1998.

Frazier, Si and Ann, "Geodes: The 'Hole' Story, Part I," *Lapidary Journal*, August 1999.

Frazier, Si and Ann, "Geodes: The 'Hole' Story, Part II," *Lapidary Journal*, September 1999.

Gilg, H.A., Morteani, G., Kostitsyn, Y., Preinfalk, C., Gatter, I., Strieder, A.J., "Genesis of Amethyst Geodes in Basaltic Rocks of the Serra Geral Formation (Ametista do Sul, Rio

Grande do Sul, Brazil): A Fluid Inclusion, REE, Oxygen, Carbon, and Isotope Study on Basalt, Quartz, and Calcite", *Mineralium Deposita*, 38: 1009-1025, 2003.

Hayes, J.B., "Geodes and Concretions from the Mississippian Warsaw Formation," *Journal of Sedimentary Petrology*, Volume LXX, 1964.

Heusser, George, "Agatized Coral at Baileys Bluff," *Gems and Minerals*, January 1983.

Juchem, Pedro Luiz, "Mineralogia, geologia e gênese dos depósitos de ametista da região do Alto Uruguai, Rio Grande do Sul," Instituto de Geoscincias, Universidade de São Paulo, 1999.

Kahrs, Margaret, Indiana's Geodized Fossils, personal communication, 2002.

Kappele, Wm., "Kentucky Fossils and Geodes," *Rock and Gem*, August 1996.

Keller, Peter C., "Geology of the Sierra del Gallego Area, Chihuahua, Mexico," University of Texas at Austin, Unpublished Ph.D. dissertation, 124 pp., 1977.

Lamb, M. & C., "The Wonderful Keokuk Geode," *Gems and Minerals*, February 1962.

Ley, Richard M., "A Notable Millerite Locality Near Bedford, Indiana," *The Mineralogical Record,* September-October 1991.

MacFall, Russell, "Florida Coral — Treasure from the Sea," *Lapidary Journal*, June 1974.

Maliva, R.G., "Model for Origin of Geodes in Ramp Creek Formations," *Geological Society of America*, No. 17, 1986.

Maliva, R.G., "Quartz Geodes – Early Diagenetic Silicified Nodules," *Journal of Sedimentary Petrology*, Volume 57, Number 6, 1987.

Maves, Sherry, "Geode Lady," *Rock and Gem*, March 1997.

Millson, Henry, "Ballast Point Agatized Coral – Part I," *Lapidary Journal*, February 1980.

Millson, Henry, "Ballast Point Agatized Coral – Part II," *Lapidary Journal*, March 1980.

Mitchell, S.M., Goodell, P.C., Lemone, D.V., Pingitore, N.E., "Uranium Mineralization of Sierra Gomez, Chihuahua, Mexico," AAPG Studies in Geology No. 13, 1981.

Morris, Alicia, "Rich in Geodes," *Rock and Gem*, March 1998.

Mozley, Peter S., "Concretions, Bombs, and Ground Water," *Lite Geology*, New Mexico Bureau of Geology and Mineral Resources, New Mexico Tech, Winter 1995.

Pabian, Roger K., and Zarins, A., "Banded Agates – Origins and Inclusions," University of Nebraska, Lincoln, Educational Circular No. 12, 32 pp., 1994.

Roder, Clara, "Tampa's Coral Wonders," *Lapidary Journal*, October 1964.

Rolm, Kenneth, "Bluegrass Geodes," *Rock and Gem*, May 1984.

Ross, C.S., and Smith R.L., "Ash-flow Tuffs – Their Origin, Geologic Relations, and Identification," U.S. Geological Survey Professional Paper No. 366, 81 pp., 1941.

Smith, J.D., "Geode Fever," *Rock and Gem*, October 1995.

Stallard, Margaret, "Florida Coral," *Rocks and Minerals*, July 1979.

Stockwell, John, "Thundereggs: Distribution and Geologic Setting," in Symposium on Agate and Cryptocrystalline Quartz, Program and Abstracts, Denver, Colorado, September 2005.

Stockwell, John, "Thinking About Thundereggs: An Historical Sketch of Inquiry into Their Nature and Origin," in Symposium on Agate and Cryptocrystalline Quartz, Program and Abstracts, Denver, Colorado, September 2005.

Tripp, Richard, "Inclusive Minerals in Keokuk Geodes," *Earth Science*, February 1961.

Tucker, M.E., "Quartz Replaced Anhydrite Nodules," *Earth Science*, November 1961.

Vaisvil, Kenneth, "Colorful Keokuk Geodes from Lewis County, Missouri," *Rocks & Minerals*, July/August 2003.

Van Tuyl, F.M., "The Geodes of the Keokuk Beds," *American Journal of Science*, Volume XLII, Fourth Series.

Vasconcelos, P.M., "40Ar/39Ar Dating of Celadonite and the Formation of Amethyst Geodes in the Paraná Continental Flood Basalt Province," American Geophysical Union, 1998 Fall Meeting, San Francisco, CA, 1998.

Wang, Y. and Merino, E., "Origin of Fibrosity and Banding in Agates from Flood Basalts," *American Journal of Science*, 295:49-77, 1995.

Wilson, Ben Hur, "Iowa's Famous Geode Park," *Earth Science*, March 1995.

Wollin, Jay, "Dig That Coral," *Rock and Gem*, October 1970.

Zeitner, June Culp, "When Geodes are Gems," *Lapidary Journal*, March 1967.

Zeitner, June Culp, "Tampa and Florida Gems and More," *Lapidary Journal*, May 1979.

MAGAZINES AND JOURNALS

Rocks and Minerals
1319 18th St. NW
Washington, DC 20036

Rock and Gem
290 Maple Ct., Suite 232
Ventura, CA 93003

Lapidary Journal
300 Chesterfield Parkway, Suite 100
Malvern, PA 19355

Colored Stone
300 Chesterfield Parkway, Suite 100
Malvern, PA 19355

Mineralogical Record
P.O. Box 35565
Tucson, Arizona 85750

Gems and Gemology
5345 Armada Dr.
Carlsbad, CA 92008

Metal, Stone, & Glass
GPO Box 1850
Brisbane, Qld., Australia

SCHOOLS

William Holland School of Lapidary Art
P.O. Box 980
Young Harris, GA 30582

Paris Junior College
Texas Institute of Jewelry Technology
2400 Clarksville St.
Paris, TX 75460

Penland School of Crafts
P.O. Box 37
Penland, NC 28765

Cranbrook Academy of Art
P.O. Box 801
Bloomfield Hills, MI 48343

Taos Institute of Arts
108B Civic Plaza
Taos, NM 87571

Colorado Academy of Silversmithing
P.O. Box 2433
Estes Park, CO 80517

Gemological Institute of America
5345 Armada
Carlsbad, CA 92008

Rochester Institute of Technology
One Lomb Memorial Dr.
Rochester, NY 14623-5603

Revere Academy of Jewelry Arts
760 Market St., Suite 900
San Francisco, CA 94102

ASSOCIATIONS

Geological Society of America
P.O. Box 9140
Boulder, CO 80301

American Society of Gemcutters
P.O. Box 826
Beaverton, OR 97004

American Federation of
Mineralogical Societies
AFMS Editor
Box 302
Glyndon, MD 21071

Friends of Mineralogy
Web Site: www.friendsofmineralogy.org

Rockhound Lapidary Hall of Fame
HCR 74, Box 21
Murdo, SD 57559

Mineralogical Society of America
1015 18th St. NW, Suite 610
Washington, DC 20096-5212

Canada Rock Association
2620 W. Broadway
Vancouver, British Columbia V6K 2G3
CANADA

Gem and Mineral Federation of Canada
Box 136
Slocan, British Columbia V0G 2C0
CANADA

BIBLIOGRAPHY

Museums with Mineral/Gem Emphasis

Harvard Mineralogical Museum
24 Oxford St.
Cambridge, MA 02138

Denver Museum of Nature & Science
2001 Colorado Blvd.
Denver, CO 80205

A.E. Seaman Mineralogical Museum
Michigan Tech University
Houghton, MI 49931

Houston Museum of Natural Science
1 Herman Circle Drive
Houston, TX 77030

Carnegie Museum of Natural History
4400 Forbes Ave.
Pittsburg, PA 15213

Natural History Museum
   of Los Angeles County
900 Exposition Blvd.
Los Angeles, CA 90007

Arizona Sonora Desert Museum
2021 N. Kinney Rd.
Tucson, AZ 85743

U.S. National Museum of Natural History
Smithsonian Institution
Washington, DC 20560-0119

Lizzadro Museum of Lapidary Art
220 Cottage Hill
Elmhurst, IL 60126

Fallbrook Gem and Mineral Museum
260 Rocky Crest Rd.
Fallbrook, CA 92028

Black Hills Museum of Natural History
P.O. Box 614
Hill City, SD 57745

The Field Museum
1400 S. Lake Shore Dr.
Chicago, IL 60605

American Museum of Natural History
Central Park West at 79th St.
New York, NY 10024

University of Arizona Mineral Museum
University of Arizona
Flandrau Science Center
Tucson, AZ 85721

Rice Northwest Museum of Rocks and
   Minerals
26385 NW Groveland Dr.
Hillsboro, OR 97124

Cincinatti Museum of Natural History and
   Science
Museum Center at Union Terminal
1301 Western Ave.
Cincinatti, OH 45203

Royal Ontario Museum
100 Queen's Park
Toronto, Ontario
M5S 2C6 CANADA

MUSEUMS WITH MINERAL/GEM EMPHASIS (CONTINUED)

Geological Survey of Canada
601 Booth St.
Ottawa, Ontario
K1A 0E8 CANADA

Guanajuato School of Mines
Lascurain de Retana No. 5
Colonia Centro
Guanajuato, Gto., Mexico 36000

Institute of Geosciences
Federal University of Minas Gerais
Av. Antonio Carlos, 6627, Pampulha
BR-31270 Belo Horizonte, M.G., BRASIL

GEOLOGICAL SURVEYS — UNITED STATES

Geological Survey of Alabama
420 Hackberry Lane
Tuscaloosa, AL 35486-6999

Alaska Division of Geological &
Geophysical Surveys
794 University Ave., Suite 2000
Fairbanks, AK 99709-3645

Arizona Geological Survey
416 W. Congress, Suite 100
Tucson, AZ 85701

Arkansas Geological Commission
3815 W. Roosevelt
Little Rock, AR 72204

California Geological Survey
801 K St., MS 12 30
Sacramento, CA 95814

Colorado Geological Survey
1313 Sherman St., Room 715
Denver, CO 80203

Connecticut Geological and Natural
History Survey
79 Elm St.
Hartford, CT 06106-5127

Delaware Geological Survey
University of Delaware
Newark, DE 19716-7501

Florida Geological Survey
Dept. of Environmental Protection
Gunter Bldg.
Florida State University Campus
903 W. Tennessee St.
Tallahasse, FL 32304-7700

Georgia Geological Survey
10 Martin Luther King, Jr. Drive,
Room 400
Atlanta, GA 30334-9004

Hawaii Commission on Water Resource
Management
Dept. of Land and Natural Resources
1151 Punchbowl St., Room 227
Honolulu, HI 96813

GEOLOGICAL SURVEYS — UNITED STATES (CONTINUED)

Idaho Geological Survey
University of Idaho
Morrill Hall, Third Floor
Moscow, ID 83843-3014

Illinois State Geological Survey
Natural Resources Bldg.
615 E. Peabody Dr.
Champaign, IL 61820

Indiana Geological Survey
611 N. Walnut Grove
Bloomington, IN 47405-2208

Iowa Geological Survey Bureau
109 Trowbridge Hall
Iowa City, IA 52242-2208

Kansas Geological Survey
University of Kansas
1930 Constant Ave.
Lawrence, KS 66047

Kentucky Geological Survey
228 Mining & Mineral Resources Bldg.
University of Kentucky
Lexington, KY 40506-0107

Louisiana Geological Survey
Louisiana State University
3079 Energy
Baton Rouge, LA 70893

Maine Geological Survey
22 State House Station
Augusta, ME 04333-0022

Maryland Geological Survey
2300 St. Paul St.
Baltimore, MD 21218

Massachusettes Geological Survey
100 Cambridge St.
Boston, MA 02202

Michigan Dept. of Environmental Quality
Geological Survey Division
P.O. Box 30256
Lansing, MI 48909

Minnesota Geological Survey
2642 University Ave. W.
St. Paul, MN 55114-1057

Mississippi Dept. of Environmental Quality
Office of Geology
P.O. Box 20307
Jackson, MS 39289-1307

Missouri Department of Natural Resources
Division of Geology
P.O. Box 250
Rolla, MO 65402-0250

Montana Bureau of Mines and Geology
1300 N. 27th St.
Billings, MT 59101

Nebraska Conservation and Survey
    Division
University of Nebraska – Lincoln
113 Nebraska Hall
Lincoln, NE 68588-0517

GEOLOGICAL SURVEYS — UNITED STATES (CONTINUED)

Nevada Bureau of Mines & Geology
Mail Stop 178
University of Nevada
Reno, NV 89577-0088

New Hampshire Dept. of Environmental
   Services
29 Hazen Dr.
P.O. Box 95
Concord, NH 03302-0095

New Mexico Bureau of Mines & Mineral
   Resources
810 Leroy Place
New Mexico Tech
Socorro, NM 87801-4796

New Jersey Geological Survey
P.O. Box 427
Trenton, NJ 08625

New York State Geological Survey
3140 Cultural Education Center
Empire State Plaza
Albany, NY 12230

North Carolina Geological Survey
1612 Mail Service Center
Raleigh, NC 27699-1612

North Dakota Geological Survey
600 E. Boulevard Ave.
Bismark, ND 58505-0840

Ohio Department of Natural Resources
Division of Geological Survey
4383 Fountain Square Dr.
Columbus, OH 43224-1362

Oklahoma Geological Survey
University of Oklahoma
Energy Center
100 E. Boyd, Suite N131
Norman, OK 73019-0628

Oregon Department of Geology & Mineral
   Industries
800 NE Oregon St., Suite 965
Portland, OR 97232-2162

Pennsylvania Bureau of Topographic and
   Geological Survey
P.O. Box 8453
Harrisburg, PA 17105-8453

Rhode Island Geological Survey
9 East Alumni Ave.
314 Woodward Hall
University of Rhode Island
Kingston, RI 02881

South Carolina Geological Survey
5 Geology Rd.
Columbia, SC 29212-4089

South Dakota Geological Survey
University of South Dakota Science
   Center
414 E. Clark St.
Vermillion, SD 57069-2390

Tennessee Dept. of Environment &
   Conservation
Geology Division
401 Church St.
13th Floor L&C Tower
Nashville, TN 37243

GEOLOGICAL SURVEYS — UNITED STATES (CONTINUED)

Texas Bureau of Economic Geology
University Station, Box X
Austin, TX 78713-8924

Utah Geological Survey
1594 W. North Temple, Ste. 3110
Salt Lake City, UT 84114-6100

Vermont Geological Survey
103 South Main
Laundry Bldg.
Waterburg, VT 05671-0301

Virginia Dept. of Mines, Minerals & Energy
P.O. Drawer 900
Big Stone Gap, VA 24219

Washington State Department of Natural
    Resources
Division of Geology, P.O. Box 47001
Olympia, WA 98504-7007

West Virginia Geological & Economic
    Survey
P.O. Box 879
Morgantown, WV 26507-0879

Wisconsin Geological and Natural History
    Survey
3817 Mineral Point Rd.
Madison, WI 53705-5100

Wyoming State Geological Survey
P.O. Box 1347
Laramie, WY 82073

GEOLOGICAL SURVEYS — CANADA

Alberta
Geological Survey of Canada
3303 – 33rd St., N.W.
Calgary, Alberta T2L 2A7

British Columbia
Geological Survey of Canada
101-605 Robson St.
Vancouver, British Columbia V6B 5J3

Geological Survey of Canada
9860 West Saanich Rd.
Sidney, British Columbia V8L 4B2

Nova Scotia
Geological Survey of Canada
1 Challenger Drive
P.O. Box 1006
Dartmouth, Nova Scotia   B2Y 4A2

Ontario
Geological Survey of Canada
601 Booth St.
Ottawa, Ontario
K1A OE8

Québec
Geological Survey of Canada
880 Chemin Sainte-Foy
Suite 840
Québec, Québec
G1S 2L2

BIBLIOGRAPHY

GEOLOGICAL SURVEYS — MEXICO

Instituto de Geologia
Universidad Nacional Autonama de
  Mexico (UNAM)
Ciudad Universitaria
C.P. 04510
Mexico City, Mexico

GEOLOGICAL SURVEYS — BRAZIL

Servicio Geológico de Brasil
Av. Pasteur, 404
Urca
22292-240
Rio de Janeiro, Brasil

GLOSSARY

Acicular - An aggregate of long, slender, crystals (i.e. Natrolite). This term is also used to describe the crystal habit of single long, thin, slender crystals.

Alkaline – A strongly basic substance. Said of a feldspar or group of feldspars containing alkali metals (e.g., sodium, potassium) but little calcium.

Alteration – Any change in the mineralogical composition of a rock brought about by physical or chemical means.

Amethyst – A pale lavender to deep purple variety of crystalline quartz.

Amygdule – A void in volcanic rock formed by steam and gases and filled by mineralization such as agate, calcite, etc.

Anhydrous – Said of a mineral that is completely or essentially without water. An anhydrous mineral contains no water in its chemical composition.

Anhydrite – Calcium Sulfate, a waterless mineral similar to gypsum.

Arch – An anticline or crest formed by a fold of strata.

Banded Agate – Agate with colors usually arranged in delicate parallel alternating bands or stripes of varying thickness. The bands are sometimes straight but usually wavy and concentric.

Banding - The presence of color zoning lines, or "bands", in some minerals.

Basalt – A fine-grained dark igneous rock which is low in silica. Basalt is usually composed of pyroxene and feldspar.

Bed – The smallest layer of sedimentary rock. It is usually a layer from one depositional event such as a flood.

Bitumen – Generic name for hydrocarbons such as tar, pitch, asphalt, and petroleum.

Botryoidal – Having the form of a bunch of grapes. Said of mineral deposits having a surface of spherical shapes. It also refers to a crystal structure in which the spherical shapes are composed of radiating crystals.

Calcite – A common rock-forming mineral consisting of crystallized calcium carbonate. It is the principal constituent of limestone; calcite also occurs crystalline in marble, loose and earthy in chalk, and stalactitic in cave deposits.

Carbonates - Group of minerals that contain one or more metallic elements plus the carbonate radical (CO_3). Most are lightly colored and transparent when pure. All carbonates are soft, brittle, and effervesce when exposed to warm hydrochloric acid.

Celadonite – A soft, green earthy mineral of the mica group, consisting of a hydrous silicate of iron, magnesium, and potassium which generally occurs in the cavities of basaltic rocks.

Chalcedony – (pronounced kal-sid-knee) A cryptocrystalline variety of quartz. It is commonly microscopically fibrous, may be translucent or semi-transparent, and has a nearly wax-like luster. Chalcedony is a catch all term that includes many well-known varieties of cryptocrystalline quartz gemstones. Chalcedony is found in all 50 States, in many colors and color combinations, and in sedimentary, igneous, and metamorphic rocks. Chalcedony includes agate, sard, plasma, prase, bloodstone, chrysoprase, flint, chert, jasper, petrified wood, and petrified dinosaur bone just to name a few of the better known varieties.

Chlorite – A group of platy and unusually greenish minerals. The chlorite minerals are usually high in iron and have an absence of calcium. They many times occur as an alteration product of iron-magnesium minerals in igneous rocks.

Citrine – A transparent yellow or yellow-brown variety of crystalline quartz closely resembling topaz in color.

Colloform – The rounded, globular texture of a colloidal mineral deposit.

Colloid – A particle-size range of less than 0.00024 mm (smaller than clay size). A fine grained material in suspension or any such material that can be easily suspended.

Concentric – Aggregate describing foliated masses that are somewhat spherical and rotate about a center; appearing like a rose (rosette). Also used to describe a form of banding where the bands are circular, forming rings about a central point.

Concretion – A hard, compact, rounded, normally sub-spherical mass or aggregate of mineral matter formed by solution often around a nucleus. Concretions normally form in the pores of sedimentary or fragmental volcanic rock and are usually of a composition widely different from that of the rock in which it is found and from which it is rather sharply separated.

Core – The central silicified portion of a thunderegg which remains after the outer rhyolite shell has weathered away. All of the core's faces are indented towards the center from which a fibrous radial structure projects.

Crinoid – An Echinoderm with long stem and a flower-like head.

Cryptocrystalline - Composed of tiny, microscopic crystals.

Crystal - Any particular three-dimensional form a mineral exhibits; which is classified by the distance ratio and angle of constituent parts. Crystals have a regularly repeating atomic arrangement that may be outwardly expressed by natural planar surfaces called "faces."

Crystallization - The forming of crystals or to assume a crystal shape.

Crystallize - To form a crystal shape, or to have crystals in a particular group (i.e. diamond *crystallizes* in the isometric system).

Dehydrated – A mineral compound in which water is not part of the chemical composition. To cause to remove water from the chemical composition of a mineral.

Dendrites – Branch-like or moss-like figures produced on or in a mineral by a foreign mineral, usually an oxide of manganese.

Deposit - An accumulation of certain minerals within a rock formation.

Devitrification - Changing over from a natural glass to a mineral with a crystalline structure.

Devitrify - The process of a natural glass to lose its glassy nature and crystallize.

Diagenetic – Indicates origin within sediment under the interface of the sea floor.

Dolomite – A carbonate sedimentary rock consisting of clay-sized dolomite crystals, interpreted as a lithified dolomite mud. Dolomite is also a common rock-forming mineral – a carbonate rich in calcium and magnesium. $CaMg(CO_3)_2$.

Dolostone – Rock primarily consisting of the mineral dolomite rather than limestone which is mostly calcite.

Double – An informal term used to describe two geodes which have been welded or joined together by nature.

Double Terminated – Referring to a mineral crystal, such as quartz, which has a termination or point on each end.

Druse – Cavity in a mineral or rock filled with protruding crystals. The hole is either completely or partially filled with crystals.

Drusy – Aggregate composed of prismatic crystals protruding from a cavity or wall.

Enhydro – Geodes which contain water.

Erosion – Process where rock is worn away from natural procedures, such as water and wind.

Exsolve – The process where an initially homogeneous solid solution separates into two or more distinct crystalline phases without addition or removal of material to or from the system. It generally occurs on cooling.

Extrusion – The volcanic process of emitting lava onto the Earth's surface.

Euhedral – Term applied to crystals of good shape.

Facet – A desired surface displayed in a gem. It may grow naturally but is usually hand cut. This definition includes the meaning of a specific cut for gems.

Faceting – Cutting from a rough stone into a gem, creating a facet.

Feldspar – Any mineral that belongs to the feldspar group. The feldspar group of minerals are aluminum silicates containing potassium, sodium, and/or calcium. This is the most abundant group of minerals on the earth, and the building block of many rocks. The feldspar group is in the tectosilicates subdivision of the silicates group.

Fibrous – Aggregate describing a mineral constructed of fine, usually parallel threads. Some fibrous minerals contain cloth-like flexibility, meaning they can be bent around and feel like cotton.

Filiform – A hair-like crystal. Also capillary.

Fluorescence – Property of certain minerals in which it displays a multicolored effect when having ample illumination with ultraviolet light.

Geode – A natural inorganic object, most often chalcedony, which are or have been hollow. Geodes are roughly spherical and can occur in igneous or sedimentary rock. The interior may be lined with crystals, including quartz as well as other minerals, pointing toward the center.

Goethite – A red, brown or yellow iron mineral and is the commonest constituent of many forms of natural rust or of limonite.

Gypsum – A soft, white mineral; hydrated calcium sulfate. Commonly formed by evaporation.

Heat treated – Said of a mineral or gem put under intense heat to enhance color or remove flaws.

Hematite – A mineral, iron oxide, constituting an important iron ore and occurring in crystals or in a red earthy form.

Hydrated – A mineral compound in which water is part of the chemical composition. To cause to incorporate water into the chemical composition of a mineral.

Igneous – A rock or mineral that solidified from molten or partly molten material, i.e., from a magma. Lavas and basalts are igneous rocks.

Impurity – An item present in a mineral which is not part of its integral structure, and may change its optical properties, such as color.

Inclusion – Materials that are locked inside a mineral as it is forming. Inclusions may also be gas bubbles or liquid-formed cavities.

Indigenous – Rocks which originated in place – where they are now found.

Lapidary – An individual who cuts gemstones as a hobby or trade, and the shop of such an individual. Also used in adjective form when relating to the cutting, shaping, and polishing of stones to exhibit their beauty and structure.

Limestone – A sedimentary rock, consisting chiefly of calcium carbonate, primarily in the form of the mineral calcite.

Limonite – A general term for a group of brown, naturally occurring hydrous iron oxides.

Macrocrystalline – Having crystals large enough to be seen with an unaided eye.

Magma – Mobile molten rock material from which igneous rocks form by solidification.

Matrix - A material that has an embedded crystal inside or emerging from it.

Metamorphic – Pertaining to the change in the mineralogical, structural, or textural composition of rocks under pressure, heat, chemical action, e.g., which turns limestone into marble, granite into gneiss.

Meteoric – Pertaining to or derived from the Earth's atmosphere.

Microcrystalline – Made up of many very small though discernible crystals.

Micromount – A mineral specimen, usually in crystal form, that requires some degree of magnification in order to be viewed.

Montmorillonite – A clay mineral complex of hydrated sodium, potassium, aluminum, and magnesium silicates.

Muck – waste rock.

Mudstone – An indurated mud having the texture and composition of shale, but lacking its fine lamination.

Nodular - Spherical, in the shape of a small rounded lump.

Nodule - Aggregate consisting of a spherical lump, usually from groups of small crystals.

Occurrence - The area where a particular mineral is found.

Opal – A mineral that is hydrated amorphous silica, softer and less dense than quartz and typically with a definite and often marked iridescent play of colors.

Osagean – A division of the Mississippian Period including the Salem, Warsaw, and Keokuk formations.

Outcrop – Rock exposed at the surface of the earth.

Oxides – Group of minerals that are compounds of one or more metallic elements combined with oxygen, water, or hydroxyl (OH). The oxide group contains the greatest variations of physical properties. Some are hard, some soft. Some have a metallic luster, others are clear and transparent.

Paramorph – Similar to a pseudomorph except that it has the same chemical composition as the original mineral.

Perlite – A hydrated, silica-rich volcanic glass.

Petrology – The science of the study of the history of rocks, their origins, physical descriptions and alterations.

Phenocryst – A large conspicuous crystal in a volcanic rock.

Pocket – Cavity in igneous rock in which crystals are usually found.

Polyhedron – A three dimensional figure composed of specific shapes.

Pseudomorph - One mineral that chemically replaces another mineral without changing the external form of the original mineral. There are three types of pseudomorphs: paramorphs, infiltration pseudomorphs, and incrustation pseudomorphs.

Pyrite – A common isometric mineral that consists of iron disulfide and has a pale brass-yellow color and metallic luster. Frequently crystallized and is also massive in mammillary forms with a fibrous or stalactite structure with a crystalline surface.

Quartz – A mineral, silicon dioxide, that occurs in colorless and transparent or colored hexagonal crystals and also in crystalline masses. An important rock forming mineral.

Quartz group – Synonym of the Silica group.

Replacement – The process of one mineral taking the place of another mineral or material, with one or two atoms per molecule in the structure being exchanged with a different one with similar characteristics, thus creating a new mineral that retains the shape of the first mineral.

Rhyolite – Fine-grained granitic rock which generally exhibits a flow texture with phenocrysts of quartz and feldspar.

Rock – An indefinite mixture of naturally occurring substances, mainly minerals. Its composition may vary in containment of minerals and organic substances, and are never exact.

Rock crystal – Transparent, colorless, crystal of quartz.

Rosette – Mineral with concentric aggregates resembling rose flowers.

Rough – Without any crystal faces. In regard to gemstones it refers to uncut material.

Sagenite – A myriad of usually tiny needle-like crystals. The actual minerals of which the needles are composed are generally not specified although the mineral rutile is one likely candidate. "Sagenite" is a descriptive term and not a scientific term.

Saponite – A soft, soapy magnesium-rich clay mineral of the montmorillonite group.

Secondary Mineral – A mineral formed after existing minerals have been altered by downward percolating waters or upwards rising gas or liquid.

Sedimentary rock – Rock formed by the weathering of substances; forming layers from accumulation of minerals and organic substances.

Selenite – A variety of sulfate of lime or gypsum occurring in transparent crystals or crystalline masses.

Shale – A fine-grained detrital sedimentary rock formed by the consolidation of clay, silt, or mud and characterized by fine, thin layers.

Shell – The outer layer of mineral matter which preserves the geodes as discrete entities. The outer layer of the geode wall. See "Wall".

Siderite – An iron carbonate mineral. Siderite is usually yellowish-brown, brownish-red, or brownish-black.

Silica – Radical of silicon and oxygen. Also term for any material composed of only silicon and oxygen (and can include water), such as quartz, chalcedony, and opal.

Silica group - Group of silicate minerals (tectosilicates) composed only of silica (silicon dioxide — SiO_2). The minerals in this group are all the varieties of quartz, chalcedony, opal, tridymite, and cristobalite (and a few rarer forms of silicon dioxide). Although opal contains water in addition to silica, it is nevertheless in the silica group. The Silica Group is also known as the *Quartz Group*.

Solid – An informal term used to describe a geode that is not hollow or has no cavity.

Spherulite – A more or less spherical body with a radial internal structure.

Thunderegg – A special kind of spherical to egg-shaped nodular mass of rhyolitic material that occurs in perlite or decomposed perlite beds and in the glassy portions of welded tuffs.

Triple – An informal term used to describe three geodes which have been welded or joined together in nature.

Tuff – Porous igneous rock composed of volcanic ash compacted together.

Tunnel – Any horizontal or shallowly sloping underground passageway. In underground mining, a horizontal or shallowly sloping passageway is referred to as an "adit." Side passages off the main adit are called "drifts," with the working faces of the mine at their ends, or in chambers along them are called "stopes."

Ultraviolet – Pertaining to ultraviolet light.

Ultraviolet light – Type of electromagnetic radiation which cannot be seen by humans.

Variety – A named specific color or other quality of a gemstone species, such as ruby for red corundum.

Vesicle – A cavity of variable shape in lava formed by the entrapment of a gas bubble during solidification of the lava.

Vitrophyre – An igneous rock having a glassy groundmass. Its composition is similar to a rhyolite.

Volcanic - Having originated from a volcano.

Volcanic rock - Rock formed as a direct result of a volcano (i.e. the solidifying of magma).

Vug - Cavity in rock that is lined with long, slender crystals. A vug forms when air pockets form in cooling magma and allow crystals to form in the hollow area.

Wall – The layers of mineral matter comprising the shell and subsequent crystallization without exposed crystal faces, more or less massive. See "Shell".

Weathered - Having gone through the process of weathering.

Weathering - The passive act of a mineral that was exposed from the earth and was chemically affected in one way or another, either by air, water, pressure, or wind.

Zoning – Internal zoning is a common feature of quartz crystals. The zones are arranged concentrically parallel to the external crystal faces and are caused by successive stages of growth. The color distribution reflects the tendency of some ions to concentrate in particular parts of the crystal.

INDEX

Note: *Italicized numbers* refer to figures and figure captions.

Index